BrightRED Results

Standard Grade BIOLOGY

Fred Thornhill, Margaret Cook, Peter Dickie,
Robert Dickson, Alastair Gentleman

First published in 2008 by:

Bright Red Publishing Ltd
6 Stafford Street
Edinburgh
EH3 7AU

A CIP record for this book is available from the British Library

ISBN 978-1-906736-13-2

With thanks to Ken Vail Graphic Design, Cambridge (layout) and Project One Publishing Solutions (copy-edit)

Cover design by Caleb Rutherford – eidetic

Illustrations by Beehive Illustration (Mark Turner) and Ken Vail Graphic Design, Cambridge.

Acknowledgements

Every effort has been made to seek all copyright holders. If any have been overlooked then Bright Red Publishing will be delighted to make the necessary arrangements.

Bright Red Publishing would like to thank the Scottish Qualifications Authority for use of Past Exam Questions. Answers do not emanate from SQA.

Contents

Revising for the Standard Grade Biology course

How does this book work?

This book covers the Standard Grade Biology syllabus in an attractive and concise format.

The Knowledge and Understanding Learning Outcomes are explained in a straightforward manner with the General and Credit Level aspects covered individually and clearly indicated. Remember, at Credit Level you are expected to know and understand the General Level facts, as well as the Credit Level ones!

Problem Solving questions account for half of the total marks in Standard Grade Biology examinations. A separate chapter of this book provides comprehensive coverage of all the Problem Solving skill areas that are assessed.

Key terms are **highlighted** throughout the text and are also explained in a Glossary at the end of the book. The Glossary is arranged into sections that correspond with the main chapters of the book.

Example questions are included throughout the book; these cover all the styles of questions found in Standard Grade examinations, to ensure you are fully prepared for your exam experience. The questions illustrate essential points for both Knowledge and Understanding and Problem Solving. Answers are given to all the questions, together with detailed explanations and guidance on how you should tackle similar questions.

The syllabus

The Standard Grade Biology course covers the following seven areas of study:

1. The Biosphere
2. The World of Plants
3. Animal Survival
4. Investigating Cells
5. The Body in Action
6. Inheritance
7. Biotechnology

Alongside these areas and throughout the course, a variety of Problem Solving skills should be developed. These are:

1. Selecting information
2. Presenting information
3. Carrying out calculations
4. Commenting on experimental procedures
5. Drawing valid conclusions
6. Making valid predictions

How you will be assessed

This course is assessed by means of **two** separate examinations. One covers Credit Level performance (Grades 1 and 2), whilst the other covers General Level performance (Grades 3 and 4).

Each paper contains questions assessing your Knowledge and Understanding (facts from the seven syllabus topics) and Problem Solving abilities. The Credit paper contains **80 marks**: 40 for Knowledge and Understanding plus 40 for Problems Solving skills. The General paper contains **100 marks**: 50 for Knowledge and Understanding plus 50 for Problem Solving skills. The Knowledge and Understanding and Problem Solving elements are marked and graded separately. You will be awarded the better of the two grades achieved in the two papers for each element.

Your overall grade

You will be given an overall grade that takes into account your performance in the written examinations together with your school-based Practical Assessment.

Investigating an ecosystem

What you should know at **General** level...

An **ecosystem** is made up of a habitat together with all of the animals and plants which live there. The plants and animals can be sampled by a variety of techniques.

Organisms which do not move can be sampled using a grid of squares called a **quadrat**.

This is put at random in the area to be studied and squares containing the organism can be counted.

Ground-living invertebrates can be sampled using a **pitfall trap**. This is usually a plastic cup buried up to its rim in soil.

Non-living conditions which can affect living organisms are known as **abiotic factors**. These include, light, temperature, moisture and pH. Living organisms are adapted to survive in particular conditions, so abiotic factors can affect the distribution of organisms. For example, some plants are adapted to low light intensities and will be found in woodlands, growing in the shade of trees. Other plants are adapted to high light intensities and will be found growing in the open.

Each abiotic factor is measured using a different technique as shown in the table below:

Abiotic factor	Technique	How used
Light	Light meter	Point the meter to the brightest part of the sample area.
Soil moisture	Moisture meter	Insert the probe into the soil and take the reading. Wipe the probe before repeating at the same depth.
Temperature	Thermometer	Leave in place until temperature remains steady.
pH	pH meter	Place the probe in soil and take the reading. Wipe the probe before repeating at the same depth.

General question 1

How would you use a named technique to estimate the abundance of dandelions on a lawn? (2)

General question 1 – Answer

Technique: *Quadrat*

Use: *You would put the quadrat down at random on the lawn and count the number of squares which had dandelions in them. You would then repeat this several times.*

*There are four points needed here for your 2 marks: the name of the technique **(quadrat)**, placing it at **random**, **counting the squares** with dandelions in them, **repeating the procedure** several times. Four correct would get 2 marks, three or two correct would get 1 mark.*

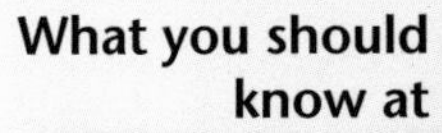

What you should know at **Credit** level...

Measurements of both the numbers of living organisms and the level of abiotic factors can give false results if not done correctly.

Common mistakes involving sampling include:

- quadrats – too few samples are taken or the quadrat is not placed at random.
- pitfall traps – the lip of the trap is above the soil surface so animals will not fall in, or the trap is not covered allowing predators to eat the animals in the trap.

Common errors made in measuring abiotic factors include:

- light intensity – letting the user's shadow fall on the light meter. This can be prevented by making sure that the meter is held out of the user's shadow.
- soil moisture – the probe could be damp or dirty. To avoid this, the probe should be wiped between each reading.
- temperature – if a thermometer is held by the bulb, the reading will not be valid. Hold the thermometer by the stem.

Credit question 1

Measuring abiotic factors can sometimes give invalid results. Identify a possible source of error for a named measurement technique and explain how it might be minimised. (1)

Credit question 1 – Answer

Measurement technique:
Measuring soil moisture.

Source of error:
The probe might be dirty or wet from the previous reading.

How to minimise it:
Wipe the probe each time before you use it.

There are several possible answers to this depending on which technique you choose to describe. Remember that you must name the technique and that the 'Source of Error' cannot be the same as the answer to how you would minimise it. For example you cannot say, 'Probe not wiped' and then say, 'Wipe the probe.' The correct way to answer it is given above.

How it works

What you should know at **General** level...

A **habitat** is the place where a plant or an animal lives.
A **population** is all of the members of one species in an area, for example, daisies in a lawn or robins in a wood.
A **community** is all the living things, both plants and animals, in an area.
All the living things in an area together with their non-living environment make up an **ecosystem**.

The size of a population can change over time. Its size depends on the birth rate and the death rate in the population. If the birth rate is higher than the death rate, the population will grow. If the death rate is higher than the birth rate, the population will decrease.

The birth and death rates can be affected by external factors which can limit the size of the population.

Examples of such factors are:
- food supply – a lack of food will prevent the population from growing.
- space – this refers to the territory a population occupies. It could restrict the population size because it affects how much food, shelter and other necessary factors are available.
- predation – a large number of animals feeding on a population will restrict its size.
- waste products – a build up of toxic waste products may increase the death rate and so reduce the population.

All living things need energy. Green plants can absorb light energy and use it to produce their own food. These plants are known as **producers**. Animals which need to feed on plants or other animals to get their energy are called **consumers**. A **predator** is an animal which feeds by killing and eating another animal. A **prey** is an animal which is killed and eaten by another animal.

A **food chain** shows the way in which energy is passed from one organism to the next. Food chains begin with plants. An example of a simple food chain is shown:

clover → rabbit → stoat

Food chains seldom exist on their own. It is much more common for them to overlap with other food chains. This results in the formation of a **food web**. An example of a food web is shown:

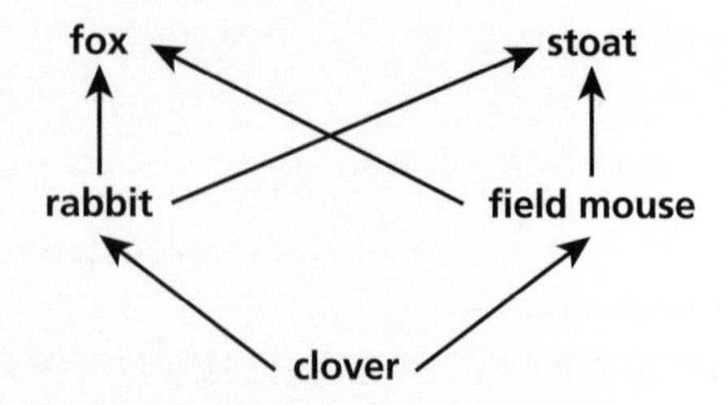

The arrows in a food chain or food web show the direction of energy flow.

Competition occurs when two organisms need the same resource. In the example above, rabbits and field mice are in competition because they both eat clover plants.

Competition can result in a reduction in the population size of both of the organisms involved. If one organism competes more successfully than the other, the less successful one might disappear from the area.

Not all of the energy in the plants at the beginning of a food chain is available to the consumers at the next stage of the food chain. This is because energy is lost at each stage of the food chain. This energy can be lost in several ways. For example:
- through movement – when an animal moves it uses up energy which is lost from the food chain.
- as heat – many chemical reactions in animals produce heat which escapes to the surroundings and again is lost from the food chain.

continued

What you should know at General level – continued

Unlike heat energy, the chemicals from which living things are made can be recycled. These **nutrient cycles** are important as they replace nutrients which have been used up in the soil.

In this way plants will not run out of the chemicals they need for growth. The chemicals in the bodies of dead animals and plants can be released during decay. The decay process which releases nutrients into the soil is carried out by micro-organisms. The nutrients can then be absorbed by plants and used for growth.

General question 1

Complete the table of words and meanings relating to the biosphere. (3)

Word	Meaning
community	All of the organisms in an area
habitat	*The place where a plant or animal lives*
ecosystem	All of the living things in an area and their non-living environment

These are three easy marks. You should have learned these definitions off by heart.

Make sure you know the difference between 'community' and 'population'.

General question 2

The following diagram represents part of a food web.

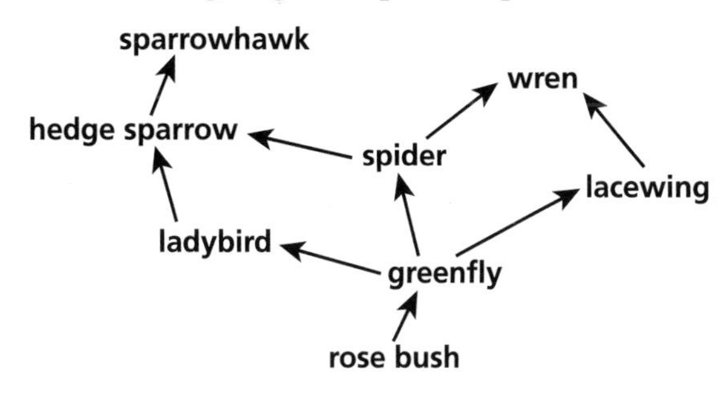

(a) What do the arrows represent in this diagram? (1)

(b) Use the food web to complete the food chain below.
rose bush → .... → .... → wren (1)

(c) Spiders and lacewings are in competition with each other. Explain what this means. (1)

(d) Most of the energy which the hedge sparrow gets from its food, does not pass on to the sparrowhawk. Give two ways in which energy can be lost. (1)

a) You must describe the arrows in terms of energy movement – nothing else will do.
b) Since there are only spaces for two organisms, you must make sure that they form part of a single food chain from the web shown.

Look out for

c) Never say that competition is when organisms fight for the same thing. Spiders and lacewings do not fight each other!

General question 2 – Answer

(a) *They show the direction in which energy moves.*

(b) rose bush → *greenfly* → *spider (or lacewing)* → wren

(c) *They both eat greenfly.* **or** *They both eat the same food.*

(d) 1 *Through movement.* 2 *As heat.*

d) You cannot use words such as 'flying' and 'walking' as two separate ways of losing energy because these are both types of movement. You must give two quite separate ways, such as movement and heat.

What you should know at Credit level...

In a food web, organisms rely on others for food. If one of the organisms is removed from a food web, those animals which used it for food will decrease in number. The species which that animal fed on, however, may increase. If two animals in the food web are in competition for a particular food and one of them is removed, the other will show an increase in its population as it now has more food.

Most animals eat prey species which are smaller than themselves. Look at the food chains below.

plant plankton → krill → blue whale

leaves → worms → blackbirds

The blue whales are much larger than the krill they eat, just as blackbirds are much larger than the worms which they eat.

The effect of this is that the number of animals in a food chain usually decreases the further you go along it. In each feeding session, a blue whale can eat over ten million krill. In turn each of these tiny shrimp-like animals might have to eat one thousand small plant plankton. This puts the number of plant plankton consumed for each whale per feeding session at about ten thousand million! If these figures are put into a diagram, we get a **pyramid of numbers**.

1 blue whale

10 000 000 krill

10 000 000 000 plant plankton

This typical pattern can be distorted if many small organisms feed on one much larger one.

Look at the food chain below.

oak tree → greenfly → blue tit → hawk

A pyramid of numbers involving these organisms may look like the one below.

1 hawk

50 blue tits

10 000 greenfly

1 oak tree

Since energy is lost at each stage of a food chain, there will be less energy available the longer the food chain becomes. The result of this is that the total mass of living material, the **biomass**, will decrease as you go along a food chain. This means that the animals at the end of the chain will have a smaller total biomass than those nearer the beginning.

This can sometimes seem confusing when a blue whale is compared to the small shrimps or krill on which it feeds. Remember, however, that since only 10% of energy is passed on at each stage in a food chain, the total mass of krill that the whale has to eat to provide its energy needs to be ten times more than the mass of the whale itself.

This can be shown in a diagram known as a **pyramid of biomass**. This is a better way of representing the organisms in a food chain.

100 tonnes of blue whale

1000 tonnes of krill

10 000 tonnes of plant plankton

continued

What you should know at Credit level – continued

The growth of a population follows a typical pattern shown in the graph opposite.

The graph shows that the population begins to increase slowly (A), followed by a rapid rise (B). Eventually the population begins to level off (C).

This pattern is caused by the initial population being small and so only able to increase slowly. As numbers increase, the population then increases rapidly with the birth rate being higher than the death rate as there is no shortage of food and there is enough space for the growing numbers. In the final stage, the birth rate and death rates are balanced as food becomes limited, waste materials might build up or space may be limited, preventing any further increase.

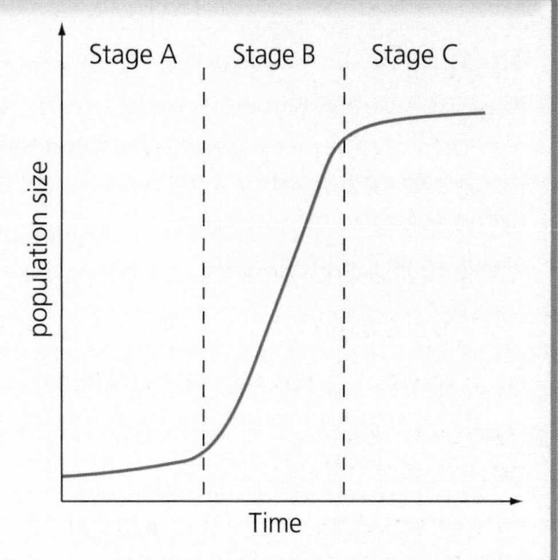

A shortage of plant nutrients in nature is avoided by nutrient cycles which recycle elements. One of the most important of these elements is nitrogen as it is needed for the production of protein in both plants and animals.

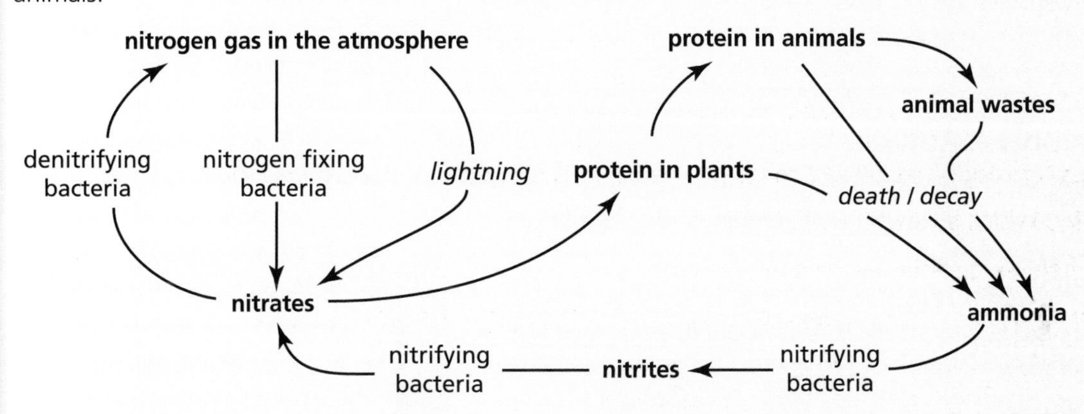

Since it is a cycle, the description can begin anywhere. The nitrogen in plant protein is passed on to animals when the plant is eaten. The dead bodies of the plants or animals or the animals' wastes decay to release ammonia.

The ammonia is acted on by soil bacteria which convert it to nitrites. Other bacteria complete the conversion of nitrites to nitrates. These are absorbed by the plant roots and used to make plant protein completing the cycle. The bacteria involved in these processes are called **nitrifying bacteria**.

Some soil bacteria act on the nitrates, breaking them down and releasing nitrogen gas into the atmosphere. These are called **denitrifying bacteria**.

Other bacteria, found mainly in nodules on the roots of plants, such as pea, bean and clover, can take in nitrogen gas and convert it into nitrates which the host plant can use. These are known as **nitrogen fixing bacteria**.

A small proportion of nitrates in the soil is formed by the action of lightning on nitrogen gas in the atmosphere.

Credit question 1

The diagram shows part of a food web from the sea.

Over-fishing has led to a decrease in the number of haddock in the food web.

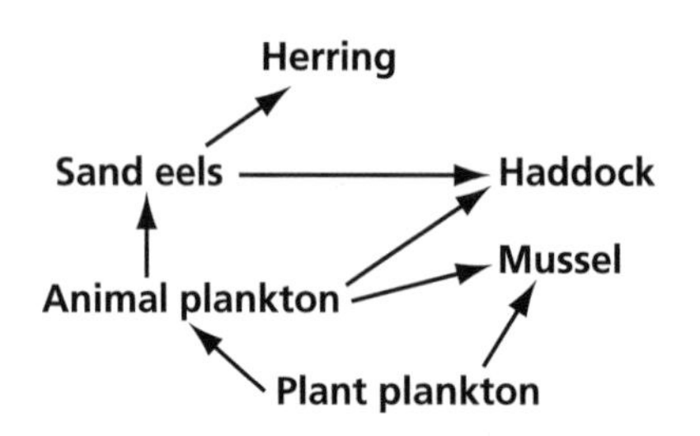

(a) Explain why the number of animal plankton might **increase** if the haddock numbers are reduced. (1)

(b) Explain why the population of animal plankton might **stay the same** if haddock numbers are reduced. (1)

(c) The following food chain is taken from the food web above.

Plant plankton → Animal plankton → Sand eels → Herring

Considering this food chain, which species would you expect to have the smallest overall mass? Give a reason for your answer. (1)

Credit question 1 – Answer

(a) *The animal plankton numbers might increase as there would be fewer haddock eating them.*

(b) *The number of animal plankton might stay the same as although there are fewer haddock eating them, there would be more sand eels which could eat them.*

(c) Species: *Herring*
Reason: *Energy is lost at each stage in the food chain so there would be the least left for the herring at the end of this chain.*

a) This is a very popular type of question. You have to look carefully at the diagram to see the relationship between the haddock and the animal plankton. From this you should be able to see that without the haddock, more animal plankton would survive.
b) From the diagram, you should be able to work out that fewer haddock would mean more sand eels which would eat more animal plankton.

Look out for

c) It is tempting to say that the smallest sized species in the food chain has the smallest mass but you must consider the *total* mass of living material. The further along the chain you go, the smaller the *total* mass at each stage becomes as energy is lost.

Credit question 2

Why are plants such as clover able to obtain nitrates even when levels of nitrates in the soil might be low? (2)

Credit question 2 – Answer

They have nitrogen fixing bacteria on their roots (1 mark)
which can take in nitrogen gas and produce nitrates. (1 mark)

Look at the mark allocation. Two marks for the question mean that you must give two facts in the answer. The first explaining about the presence of the nitrogen fixing bacteria and the second describing what these bacteria do.

Control and management

What you should know at **General** level...

All parts of the environment, air, land, fresh water and the sea, can be affected by **pollution**.

The main sources of pollution and examples of a pollutant from each are shown in the table.

Type of pollution	Typical pollutant
Domestic pollution from houses	Litter, untreated sewage, thoughtless rubbish tipping
Agricultural pollution from farming	Pesticides or excess fertilisers
Industrial pollution from factories	Smoke, oil or other chemical spills

Pollution can be reduced in many ways such as using cleaner fuels in factories or providing improved facilities for the disposal of litter and household waste. The use of unleaded petrol is another example of a way in which pollution can be reduced.

Organic waste such as untreated **sewage** or some waste from food processing plants, can provide a food source for micro-organisms such as bacteria. Increased numbers of bacteria in a river or loch use up more oxygen so that there might not be enough for other organisms such as some insects or fish.

We rely on our environment for everything, for our food and the raw materials for industry. Sometimes the way we use these natural resources can produce problems.

Poor management of the fishing industry has resulted in overfishing in the North Sea. The effect of this has been a fall in the numbers of some species such as cod. Controlling the number of fish which boats can catch can help the fish populations recover.

In some countries, overgrazing by farm animals means that not enough vegetation is left to prevent the soil from being blown away by the wind. This can be helped by planting areas of trees to stabilise the soil.

General question 1

The table shows examples of pollution of four different ecosystems.

(a) Complete the table by filling in the empty boxes. (3)

General question 1 – Answer

Ecosystem affected	*Source of pollution*	*Example of pollutant*
Air	Domestic	CFC gases from aerosol sprays
Fresh water	*Agricultural*	Pesticides in a river
Sea	Industrial	Crude oil from tanker vessels
Land	Domestic	*Uncontrolled waste dumping*

*The question asks for four **different** ecosystems which is why you must put in 'sea' not just 'water' as the missing ecosystem in line three of the table. Remember that 'waste' on its own is not enough for a domestic pollutant as all houses produce waste and it is only when it is badly handled that it becomes a pollutant.*

General question 2

The list below contains statements about pollution.

X Smoke from coal fired power stations causes acid rain.
Y Raw sewage in rivers leads to the death of fish.
Z Car exhaust fumes contain poisonous gases.

Choose **one** of the statements and give an example of a way in which the pollution **could** be controlled. (1)

Statement letter	Method of control
X	*use cleaner fuels / remove acid gases from smoke*
or *Y*	*improve sewage treatment*
or *Z*	*use unleaded petrol* **or** *reduce car use* **or** *fit catalytic converter*

Make sure that the control method is appropriate for the pollution statement you choose. The question asks for you to select a letter to identify the statement. Use the letter – do not waste your time copying out the statement description!

What you should know at Credit level...

Using fuels as energy sources causes pollution. **Fossil fuels** such as coal produce considerable quantities of smoke and sulphur dioxide gas when they burn. In addition, coal fired power stations can produce large quantities of ash which has to be disposed of.

Nuclear fuels produce much less waste but this waste is so radioactive that it will have to be stored securely for many generations.

If organic waste like raw sewage enters a river or loch, bacteria in the water will use it for food. Their numbers can then increase rapidly to very high levels. As their numbers increase, they will use up more and more oxygen. This results in a decrease in the dissolved oxygen which is available for other creatures in the water. Those with a high oxygen requirement might disappear from the water and so the total number of species will decrease. Those which are left and can tolerate lower oxygen, often multiply rapidly as competition and predation are reduced.

Since different organisms can survive in different levels of oxygen, these species can be used to predict what the oxygen level of the water is likely to be. Stoneflies, for example, need very high oxygen levels and so if they are found in the water, the oxygen level is likely to be very high. Animals are not the only organisms which can give clues about their habitat. Nettles grow particularly well on soil with a high phosphate concentration. Organisms like this are known as **indicator species** and can be used to predict the level of an environmental factor.

We often have to control various components of an ecosystem when trying to grow crops. In farming, for example, competition is reduced by using weedkillers, water might have to be provided by irrigation and nutrients are supplied to the crops by using fertilisers.

Credit question 1

The Biochemical Oxygen Demand (BOD) is a measure of how much oxygen is used up by micro-organisms in water. The higher the BOD, the more oxygen is being used up.

(a) Explain why water contaminated with raw sewage will have a high BOD. (2)

(b) What would be the effect on the species in a river if the BOD of the water increased? (1)

Credit question 1 – Answer

a) *There will be more food for the micro-organisms so they will increase in number.* (1 mark) *The increased number of micro-organisms will use up more oxygen during respiration.* (1 mark)

b) *The variety of organisms in the water would be reduced* **or** *There would be fewer species present.*

There are two marks for this question so one fact will not get full marks.
The important points are:
1 – increased food supply leading to increased numbers of micro-organisms
2 – the resulting greater oxygen uptake.

This is another example of where it is important that you read the whole question. You are told at the beginning of the question that an increased BOD means more oxygen is used up. This means that the water will have less oxygen in it leading to fewer species being able to survive. Do not say, 'There will be fewer organisms'. This is not always the case. With reduced competition, the number of bloodworms, for example, which can tolerate low oxygen levels, is often many times greater than the original number of organisms found in the area.

Credit question 2

What is meant by the term 'indicator species'? (1)

Credit question 2 – Answer

This is an organism whose presence or absence gives information about abiotic conditions in an area.

It is important when answering questions of this type that you do not reuse the words in the question itself. For example, 'It is a species which indicates things in the environment' would not get you the mark.

Introducing plants

What you should know at **General** level...

There is a great variety of different plants.

Plants have many uses:

- they provide food for people and animals
- some plants provide habitats for other plants and animals
- some plants are a source of medicines
- some plants are used to produce raw materials
- some plants are grown for their attractive appearance in homes and gardens

The great variety of different plants means that there is a wide range of plants for each use.

Examples of specialised uses of plants include:

Food	potato plants for potatoes	rice plants for rice
Medicines	opium poppy for painkillers	foxglove for heart medicine
Raw materials	pine trees for timber	rubber trees for rubber

General question 1

Six plants used by people are named below.

tomato plant   pine tree   cotton plant
apple tree   opium poppy   foxglove

Complete the table by choosing **two** plants from the list and giving one **different** use for each plant. (2)

General question 1 – Answer

Plant	*Use*
opium poppy	*pain killer*
pine tree	*timber*

There are several possible answers to this question so choose plants that you are familiar with and know what they are used for. The important point is that the two plants you choose must have different uses.
For example:
one food plant + one medicinal plant
***or** one food plant + one raw material plant*
***or** one medicinal plant + one raw material plant*
One mark given for each correct plant + use.

What you should know at **Credit** level...

Some plant species may become extinct because of the effects of factors such as climate change, deforestation and soil erosion. The loss of plant species can affect people because it reduces the range of plants available as sources of food, medicines and raw materials. Plant scientists continue to study plants to try to find potential new uses for them.

It is important to maintain as wide a variety of plants as possible because they may provide new sources of foods and medicines. The loss of plant species can affect other animals because they may lose sources of food or habitat.

The use of many plants requires a specialist production process, such as the use of barley in the production of beer. Barley grains contain stored starch which must be converted into maltose sugar before yeast can ferment the sugar into alcohol. The barley grains are allowed to germinate for a few days. During this time, enzymes in the barley change the starch into maltose. This is called **malting**.

Credit question 1

The following list contains some uses of plants used in production processes.

- Barley grains used to make beer.
- Rape seeds used as a source of oil.
- Raspberries used to make jam.
- Pine trees grown to produce timber.

Select **one** of the plants from the list and describe **one** stage in the production process. (1)

Credit question 1 – Answer

Plant: *Barley*

Description of stage in production process: *Barley grains are allowed to germinate so that the starch is changed into maltose sugar.*

The malting of barley is a good choice to make because you need to know this particular example for Topic 7 Biotechnology

Growing plants

What you should know at **General** level...

The seeds of all flowering plants have three main parts.

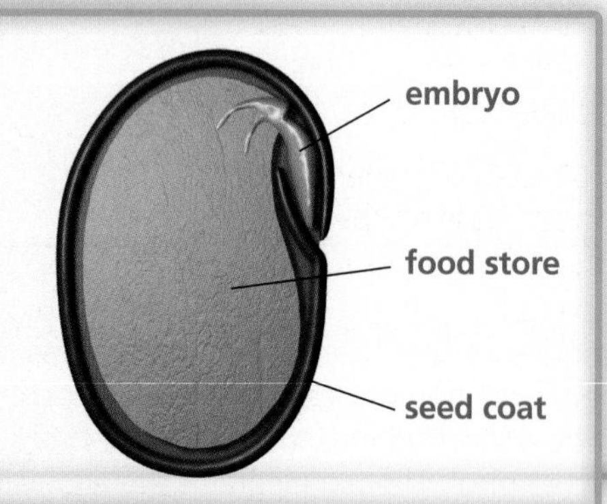

The early stage of growth of a seed is called **germination**. Successful germination needs:
- water
- oxygen
- a suitable temperature.

Seed formation is part of the **sexual reproduction** process in flowering plants.

The sequence of events involved is:

1 flowering	the production of the flowers
2 pollination	the transfer of pollen from anther to stigma
3 fertilisation	the joining of the pollen nucleus with the ovule nucleus
4 seed formation	the fertilised ovule develops into a seed
5 fruit formation	the ovary with the seeds develops into a fruit

A **pollen grain** contains the male **gamete** of the plant. The ovule contains the female gamete.

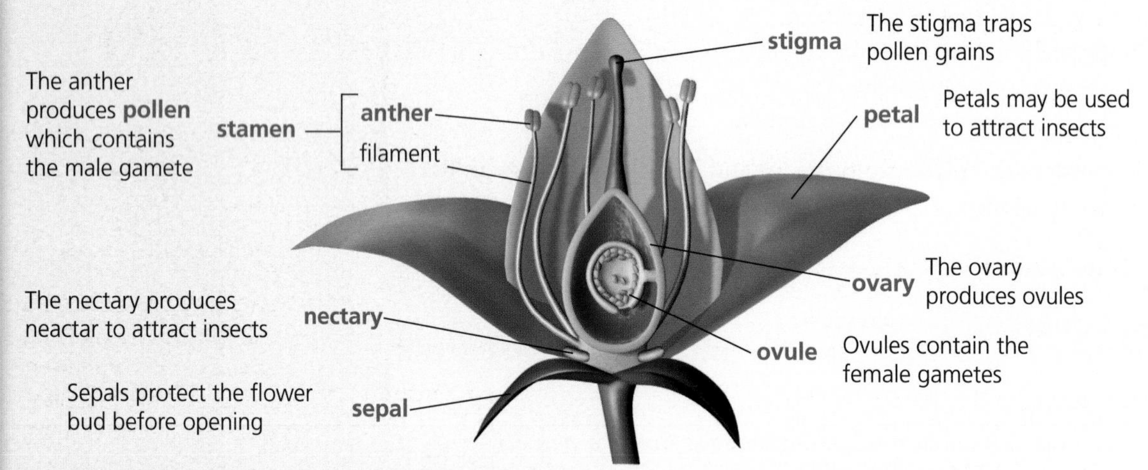

There are two common methods by which pollination takes place.

Insect pollination – Insects such as bees visit flowers to collect nectar as food. Pollen grains are brushed from the anther and stick to the insect's body. At the next flower some of the pollen from the insect sticks to the stigma.

Wind pollination – Very small and light pollen grains are blown from the anthers by the wind. Some pollen in the wind is trapped by the stigmas of other flowers.

After pollination, fertilisation can take place. This happens when the nucleus of a pollen grain reaches and joins with the nucleus of an ovule. After fertilisation, the fertilised ovule develops into a seed. The ovary containing the developing seeds becomes the fruit of the plant.

Many flowering plants can also reproduce asexually. **Asexual reproduction** does not involve fertilisation. Instead, parts of a parent plant develop and separate from the parent to become independent new plants.

continued

What you should know at General level – continued

Two examples of ways in which this happens naturally are **runners** and **tubers**.

Runners are horizontal stems which grow over the surface of the soil. Buds on the stems develop roots and leaves. The connecting stem breaks down leaving the new plant separate from its parent.

Tubers are swollen parts of roots or underground stems containing stored food. Buds on the tubers grow into new plants in the following year.

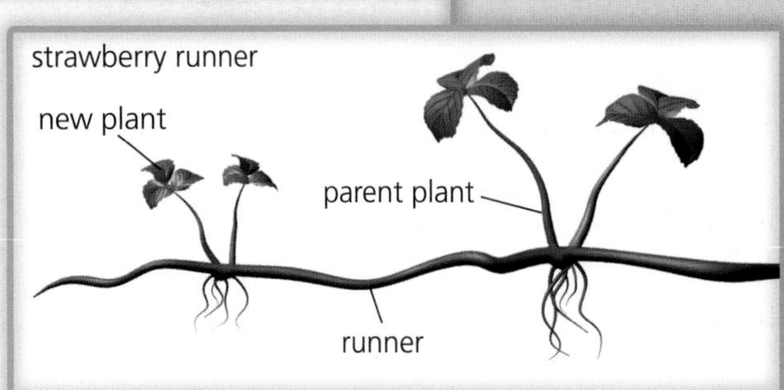

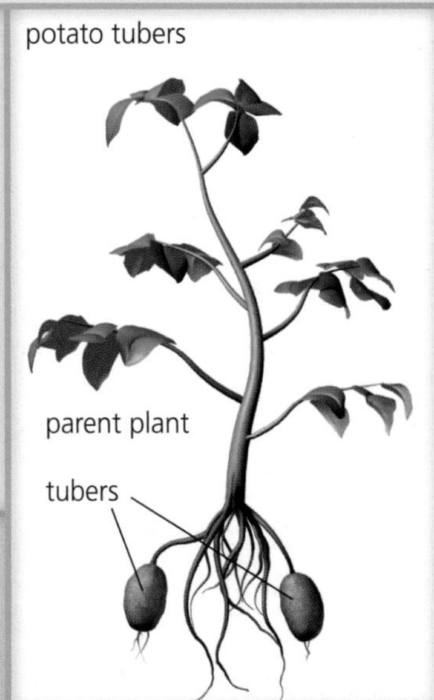

People have developed artificial methods of propagating (making more) plants asexually.

Two common ways of doing this are using cuttings and grafting.

Cuttings are sections of a stem cut from a parent plant. The cuttings are placed in soil and they grow roots and leaves. Each cutting becomes a new plant.

Grafting uses a section of a stem from a parent plant which is grafted (joined) to the roots of a stronger variety. The graft grows and shows the features of the variety it was taken from. Many grafts can be taken from the parent plant.

General question 1

Which of the following factors are needed by all seeds for germination? (1)

carbon dioxide oxygen light water

General question 1 – Answer

oxygen water

Look out for

Germination does not involve photosynthesis and so light and carbon dioxide are not needed. A third essential factor is a suitable temperature.

General question 2

Name two possible methods of pollination used by plants. (1)

General question 2 – Answer

1: *wind*

2: *insects*

Make sure you don't mix up pollination and seed dispersal. If you mention anything in this answer (such as hooks or wings) which suggests confusion between the two processes, you will lose the mark.

What you should know at Credit level...

The percentage germination of seeds changes at different temperatures. At very low temperatures (below 5°C), percentage germination is nil or very low. The percentage germination increases as temperatures increase. The highest percentage germination occurs at the optimum temperature. This differs for different plants. At temperatures above the optimum, percentage germination decreases because enzymes are denatured (see Topic 4)

The structure of a flower shows adaptations for the method of pollination used. These are shown in the diagrams below.

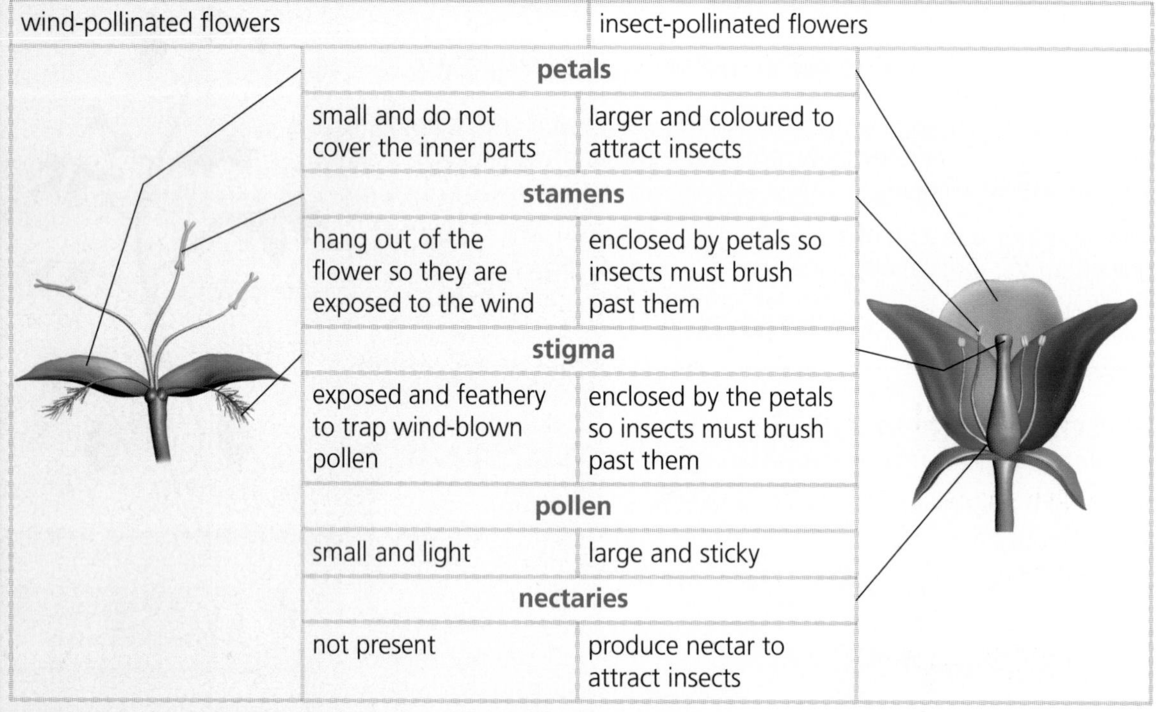

wind-pollinated flowers	insect-pollinated flowers
petals	
small and do not cover the inner parts	larger and coloured to attract insects
stamens	
hang out of the flower so they are exposed to the wind	enclosed by petals so insects must brush past them
stigma	
exposed and feathery to trap wind-blown pollen	enclosed by the petals so insects must brush past them
pollen	
small and light	large and sticky
nectaries	
not present	produce nectar to attract insects

After pollination, fertilisation happens the same way in all flowers

- A pollen grain on the stigma grows a tube down the stigma into the ovary and reaches an ovule.
- The nucleus of the pollen grain (male gamete) passes down the tube and joins with the ovule nucleus (female gamete). This is fertilisation.
- The fertilised ovule develops into a seed.
- The ovary, containing the seeds, develops into the fruit of the plant.

The fruit is adapted to help disperse the seeds away from the parent plant. This happens in three common ways, wind dispersal, animal internal dispersal and animal external dispersal.

Wind dispersal – Fruits have fine hairs or wing structures which allow them to be blown by the wind.

Animal internal dispersal – Fruits are fleshy and eaten by animals. The seeds are not digested and pass out of the animal with their faeces.

continued

What you should know at Credit level – continued

Animal external dispersal – Fruits have hooks which cling to the coats of animals and are brushed off later.

goosegrass

burdock

All asexual methods of reproduction (runners, tubers, cuttings, grafting) produce offspring which are genetically identical to each other. This means that they will all show the same characteristics. A group of genetically identical offspring produced by asexual reproduction is called a **clone**.

Artificial propagation of plants is used to produce large numbers of particular varieties by asexual methods. This is important commercially.

Both sexual and asexual reproduction have advantages for plants.

Sexual reproduction:
- causes genetic variation in the offspring allowing adaptation to changing conditions
- involves dispersal of the seeds which reduces competition between offspring.

Asexual reproduction:
- results in genetically identical offspring which means that all the characteristics of the parent will be passed to all offspring, giving them equal chances of survival
- is rapid and does not involve vulnerable stages of growth.

Credit question 1

The diagram represents the flower of a plant.

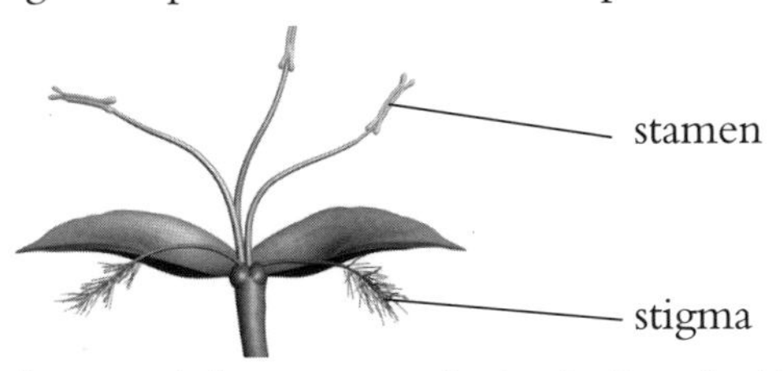

Explain how each feature contributes to the plant's method of pollination. (2)

Credit question 1 – Answer

Stigma Feature: *feathery* **or** *hangs outside of petals*
Explanation: *allows it to trap wind-blown pollen*

Stamen Feature: *hangs outside of petals*
Explanation: *allows winds to blow pollen away*

Look out for

You must be able to recognise wind-pollinated and insect-pollinated flowers. You must also remember the function of each of the structures.

Credit question 2

The following statements refer to the stages that occur after pollination.

A Fertilisation takes place.
B A pollen tube grows out from a pollen grain.
C The ovule forms a seed and the ovary forms a fruit.
D The pollen tube grows down through the stigma.
E The male gamete moves towards the ovule.
F The pollen tube grows through the ovary wall.

Use the letters of the statements to complete the sequence of stages. (2)

Credit question 2 – Answer

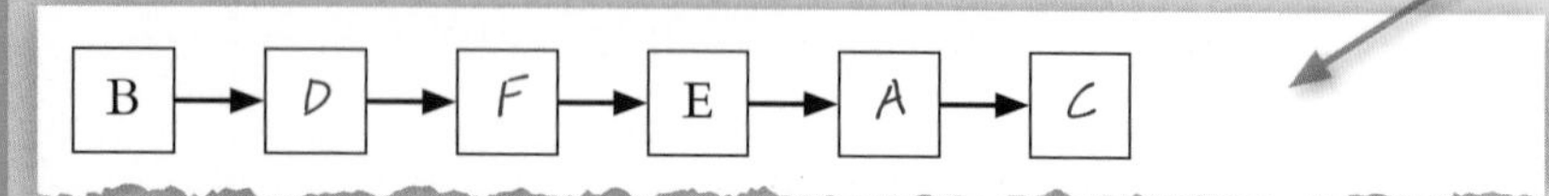

Use the boxes completed for you as guides. For example, **B** *is given as the first answer. It is logical that continued growth of the pollen tube is the next stage. In the same way, fertilisation* **(A)** *must come later than the movement of the male gamete.*

Credit question 3

The plants in a clone have been produced by asexual reproduction. Give **one** other piece of information about the members of a clone. (1)

Credit question 3 – Answer

They are genetically identical

You are asked for one fact. If you are unsure, don't try to play safe by giving two facts in case one is wrong, you will lose the mark. Think carefully and make a decision.

Making food

What you should know at General level...

Transport systems in plants are used to supply the cells with food and water.

There are two types of transport tissue.

- **Xylem** – this carries water upwards from the roots to all other parts of the plant.
- **Phloem** – this carries food such as sugar around the plant. It may be from the leaves where it was made to other areas for storage or to be used for growth.

continued

What you should know at General level – continued

Plants make sugar by the process of **photosynthesis**. Photosynthesis takes place in the leaves. The features of photosynthesis are shown in the diagram.

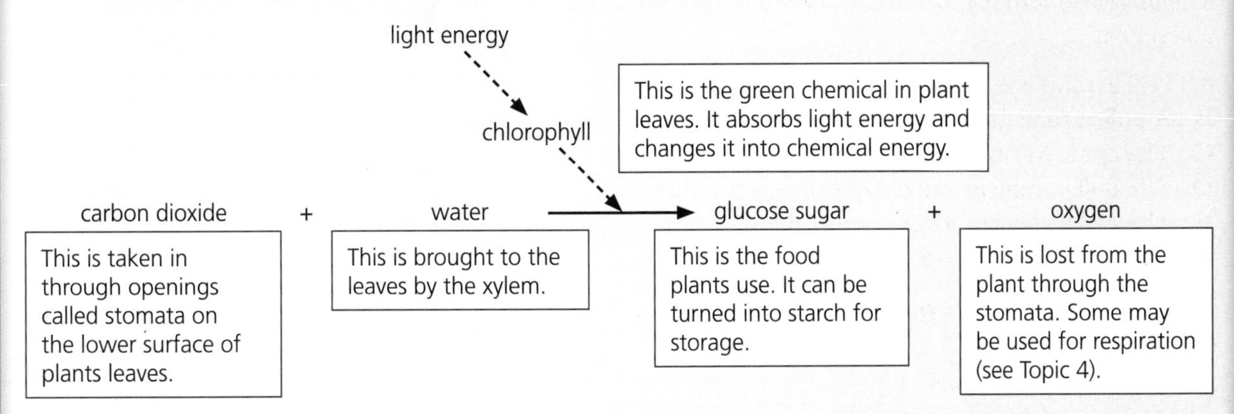

Photosynthesis depends on light energy. **Chlorophyll** in the leaves absorbs light and changes it into the chemical energy of **glucose**. The chlorophyll is present in structures called **chloroplasts**.

Plants lose water vapour through their **stomata**. The stomata close at night to prevent excessive water loss.

General question 1

Complete the word equation for photosynthesis. (2)

General question 1 – Answer

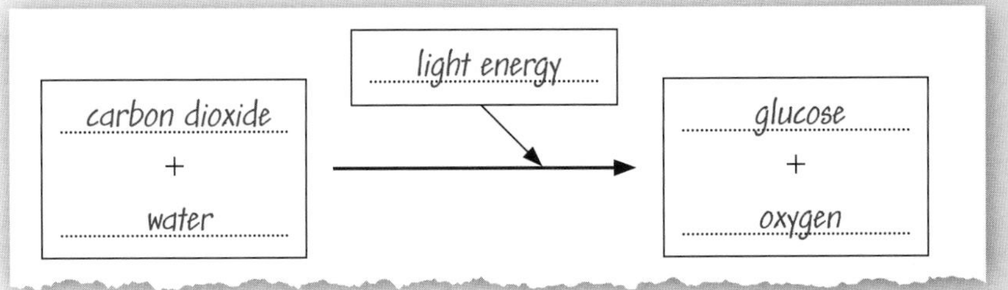

Look out for

Glucose is made during photosynthesis and so it is one of the products. It is a carbohydrate so the raw materials must provide the carbon, hydrogen and oxygen needed to make it. These are present in carbon dioxide and water so these are the raw materials. Oxygen is the other product.

General question 2

Plants exchange gases with the air during photosynthesis. Name the openings which allow gases to pass into and out of the leaf. (1)

General question 2 – Answer

stomata

Look out for

Stomata is the plural and is used when talking of more than one of these openings. A single opening is called a **stoma**.

General question 3

What substance in green leaves absorbs the light energy needed for photosynthesis? (1)

General question 3 – Answer

chlorophyll

Be careful when spelling this word. And remember: chlorophyll is the green chemical. It is present in structures called chloroplasts. Don't confuse the two words.

What you should know at Credit level...

The features of xylem and phloem are shown below.

xylem vessels		phloem vessels	
features	**diagram**	**diagram**	**features**
continuous hollow tubes formed from dead cells joined together			living cells with no nucleus, joined together into long vessels
rings of **lignin** which strengthen the vessels			**companion cell**
			end cell walls perforated to form **sieve plates**

The xylem transports minerals, as well as water, from the roots to the rest of the plant.

The strong xylem vessels also help to support the plant.

Plant leaves are adapted to carry out photosynthesis. They are flat and thin to provide a large surface area for absorbing light.

Features of a typical leaf are shown in the diagram.

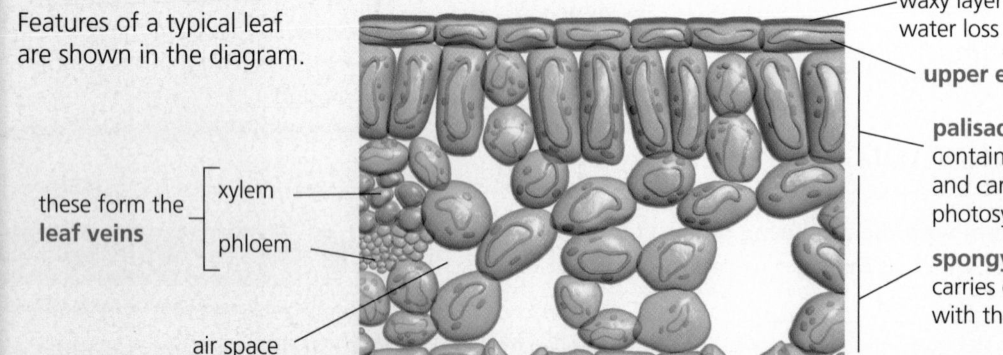

continued

What you should know at Credit level – continued

The carbon dioxide absorbed by the plant is used to make glucose sugar during photosynthesis. Plants use the glucose in different ways:
- used for respiration to release energy (see Topic 4)
- used to make the **cellulose** of plant cell walls, and other structural chemicals
- used to make **starch** which is stored for later use.

The rate of photosynthesis in a plant is affected by three main factors:
- light intensity
- carbon dioxide concentration
- temperature.

Any one of these can restrict how quickly photosynthesis takes place. For example, even in bright light and a high carbon dioxide concentration photosynthesis may be very slow if the temperature is too low for efficient enzyme activity. In this case temperature is the limiting factor for photosynthesis.

In similar ways, low light intensities or low carbon dioxide concentrations can act as **limiting factors**.

At any one time there will be one factor which is limiting the rate of photosynthesis.

Credit question 1

Which of the following comparisons between xylem and phloem is correct?
Tick the correct box. (1)

Credit question 1 – Answer

Xylem	**Phloem**	
Contains sieve plates	No sieve plates	
Made up of dead cells	Made up of living cells	
Lignin present	No lignin present	✓
Carries nutrients to roots	Carries nutrients to leaves	

*To answer this question you must think carefully about the **difference in structure between** xylem and phloem, and the **difference in their function**.*

Credit question 2

The diagram shows a magnified view of the structure of a leaf.

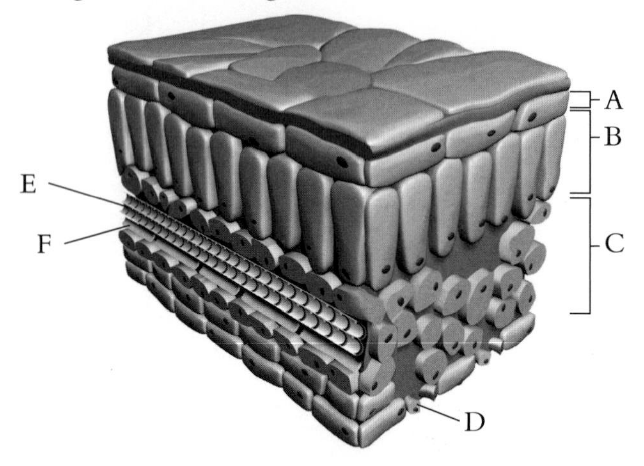

Complete the following table that describes some features of the leaf. (3)

Credit question 2 – Answer

Letter	*Name*	*Function*
A	*Epidermis*	Cells that form upper surface of the leaf
B	Palisade mesophyll	*Carries out photosynthesis*
C	*Spongy mesophyll*	Exchanges gases between air and leaf cells
D	*Guard cell*	Controls the size of the stoma
E	Xylem	*Carries water to the leaf*
F	*Phloem*	Transports glucose from the leaf

This style of question means that you need thorough knowledge of the topic in order to gain full marks. Proper revision is needed to learn these facts.

Credit question 3

The rate of photosynthesis can be limited by different factors. Draw one line from each set of conditions to the factor that would be limiting photosynthesis. (?)

Credit question 3 – Answer

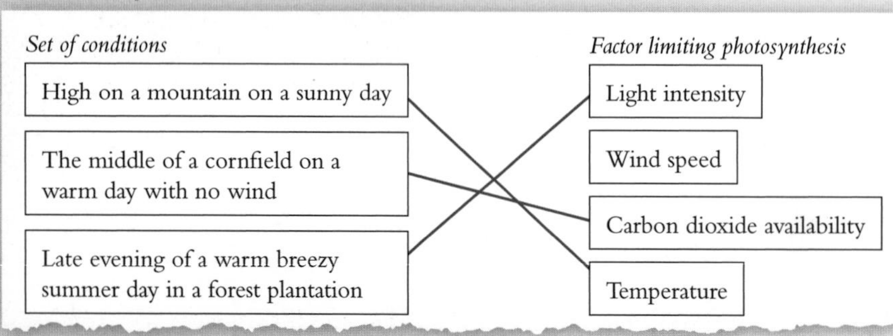

Look out for

There can be only one limiting factor in any given set of conditions. Wind speed is never a limiting factor but lack of wind can mean that the air can become short of carbon dioxide if photosynthesis has been happening rapidly.

The need for food

What you should know at **General** level...

There are 3 main types of food; carbohydrates, proteins and fats

- carbohydrates and fats are both required as sources of energy
- proteins are needed for growth and for repair of damaged or worn out tissues.

Digestion is the process of breaking down large food particles into small, soluble particles.

This means that they are small enough to pass through the wall of the small intestine and to dissolve in the blood.

Different mammals eat different types of foods:

- **herbivores** eat only plant materials; sheep and cattle are herbivores.
- **carnivores** eat only animal materials; dogs and tigers are carnivores.
- **omnivores** eat both plant and animal materials; humans are omnivores

There are only four types of teeth in mammals: incisors; canines; premolars and molars. These are used differently by different mammals, depending upon their diet. The differences are shown in the table.

	teeth		
animals	**incisors**	**canines**	**premolars and molars**
herbivore (such as sheep) 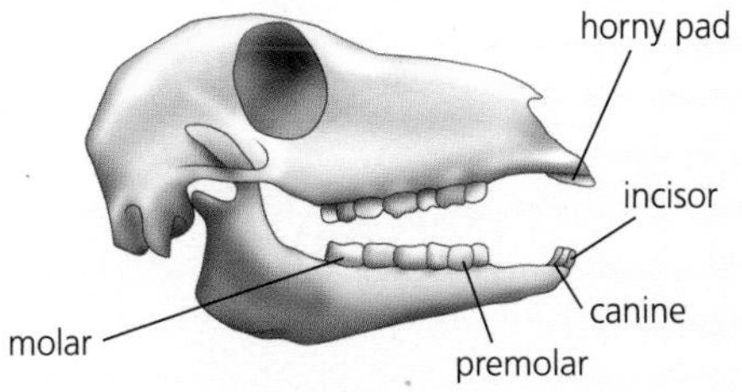 horny pad, incisor, canine, molar, premolar	These are not present on the top jaw of sheep. Lower incisors have sharp edges which cut grass against the horny pad on the top jaw.	These are not present on the top jaw. The lower canines look and act like the incisors.	These have ridges which fit close together to grind grass.
carnivore (such as dog) 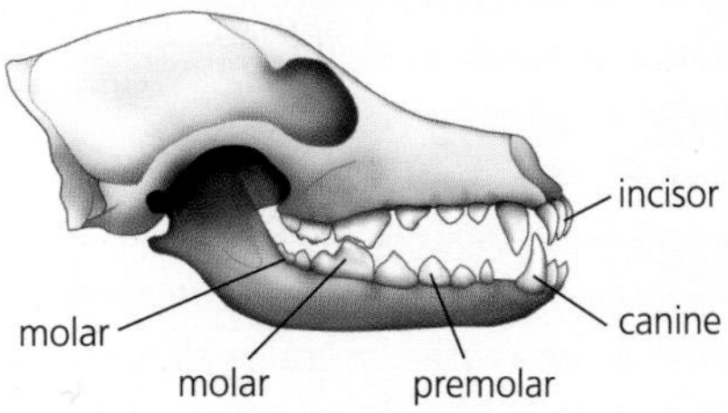 incisor, canine, molar, molar, premolar	These are small sharp teeth which pull and scrape meat from bones.	These are long and pointed to kill prey. They are then used to grip and tear the flesh.	These have sharp edges which slice flesh.
omnivore (such as human) 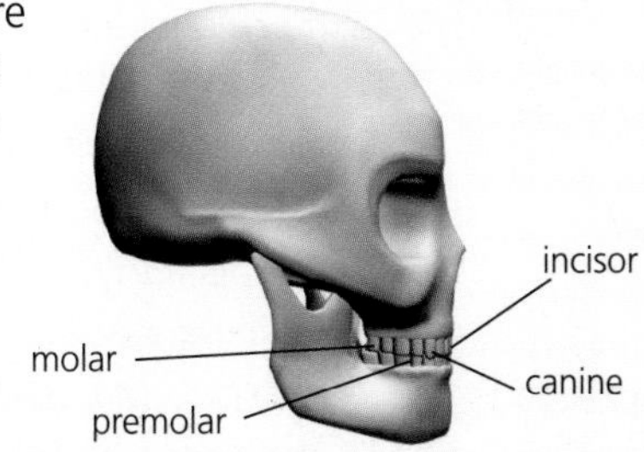 incisor, molar, canine, premolar	These have straight sharp edges for cutting food.	These are pointed and used to grip tough food.	These are wide teeth which crush food between them.

continued

What you should know at General level – continued

You must be able to identify and name the parts of the alimentary canal on the diagram together with their functions as shown.

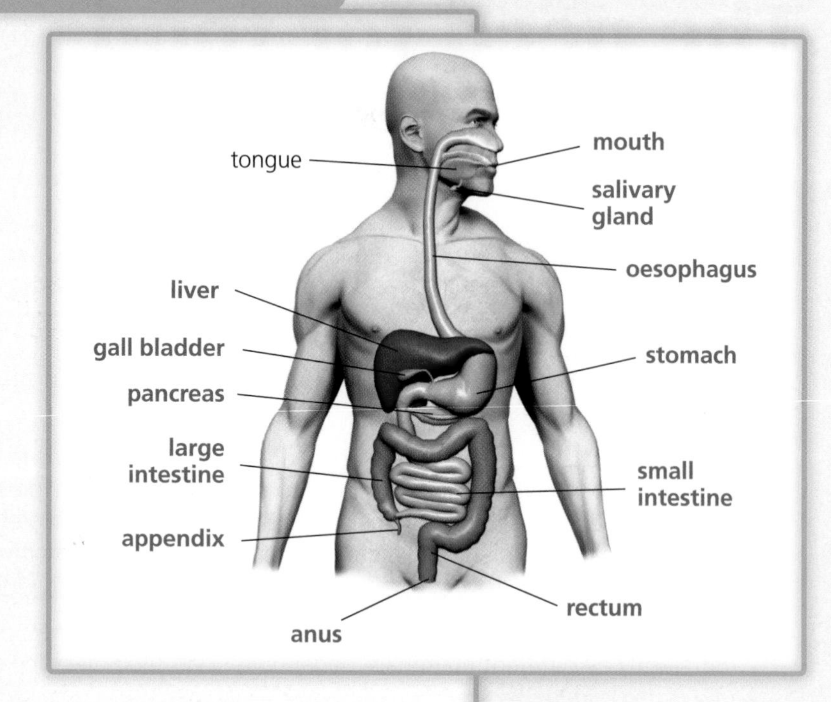

Enzymes are chemicals which speed up biological reactions. Digestion involves several enzymes which speed up the breakdown of food:

- **Protease** enzymes are used to break down proteins.
- **Amylase** enzymes are used to break down carbohydrates.
- **Lipase** enzymes are used to break down fats.

The small intestine is long and its inner surface is covered with microscopic projections called **villi**. These features mean that it has a large surface area for the absorption of digested foods.

The large intestine absorbs water from undigested food to leave faeces which passes to the rectum where it is stored until it is eliminated through the anus.

General question 1

The diagram shows the skulls of two mammals.

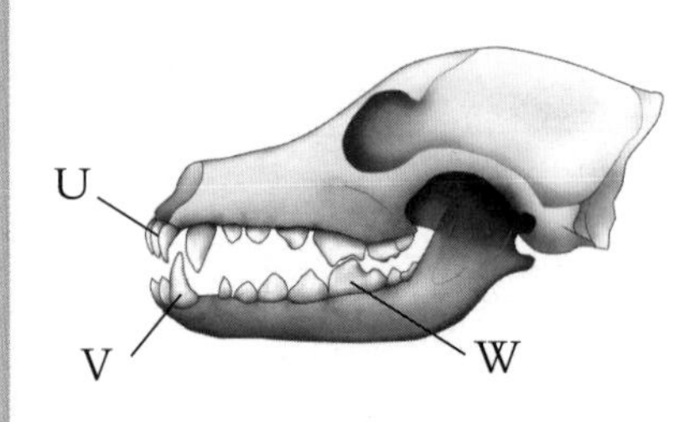

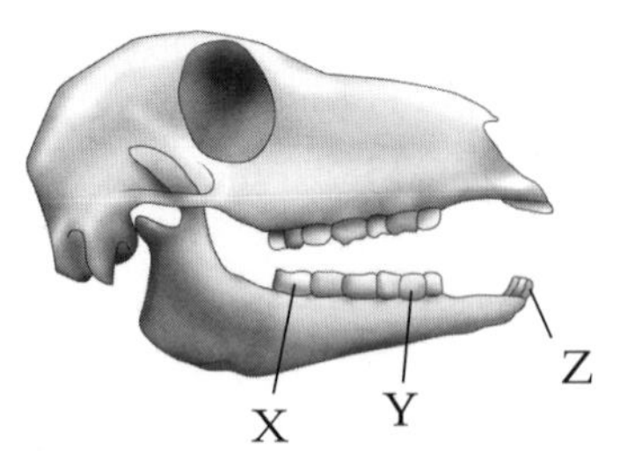

Use letters from the diagram to identify the following teeth. (3)

General question 1 – Answer

(i) Incisors: U and Z
(ii) A tooth used for piercing and holding prey: V
(iii) A tooth used for crushing and grinding plant material: X

When you look at the skulls you will automatically realise that the one on the left is a carnivore and the one on the right is a herbivore.
(i) The incisors of mammals are always nearest the front of the mouth, so ***U*** *and* ***Z****.*
(ii) Careful here – only the carnivore will hold prey, and it is the canine teeth which pierce and hold the prey, so only ***V*** *will do.*
(iii) Similarly, only the herbivore will have special teeth for plant material, and crushing and grinding is the job of the molars at the back of the mouth, so ***X****.*

General question 2

The diagrams show three different types of human teeth.

A 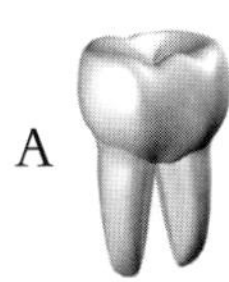 B 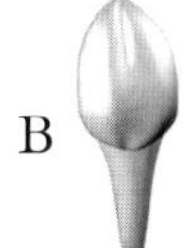 C 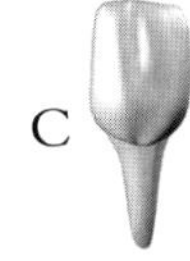

Complete the following table by choosing one of the teeth, **A**, **B** or **C**, for each description.
Each letter may be used once, more than once or not at all. (2)

General question 2 – Answer

Description of tooth	*Tooth*
Found at the very back of the jaw	A
Known as an incisor	C
Used for grinding and crushing food	A

Beware of making assumptions. Just because there are three teeth in the diagram and three descriptions don't assume that there will be one for each tooth. The sentence 'Each letter may be used once, more than once or not at all.' is a helpful hint. The only way is to answer each question from your knowledge and not from ideas of what you might expect.
The molars are the big ones at the back so the first answer must be **A**. The incisors are flat, cutting teeth so this must be **C**. Crushing and grinding is the job of the molars so **A** again.

General question 3

The diagram below shows some of the parts of the human digestive system.

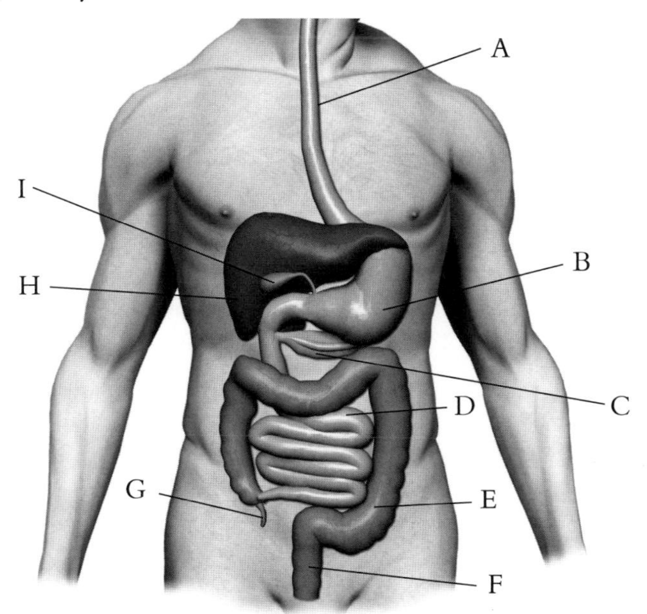

a) Complete the table overleaf using letters to identify the named parts of the digestive system. (2)
b) Name the parts labelled with the letters shown. (2)
c) What feature of the small intestine, shown on the diagram, helps it to carry out its function efficiently? (1)
d) State one function of the large intestine. (1)

General question 3 – Answer

a)

Part of digestive system	*Letter*
Gall bladder	I
Rectum	F
Stomach	B
Small intestine	D

b) A *oesophagus*
C *pancreas*
E *large intestine*
G *appendix*

c) *The small intestine is very long.*

d) *It absorbs water (from the undigested material into the blood).*

There are lots of marks for simply learning the names of structures. Thorough learning of the basic facts is well rewarded throughout the exam.

In the answer to (d) the part in brackets is optional – to make it easier to read the answer. You would get the mark for the first bit on its own. Don't worry too much about the spelling of complicated words like oesophagus. If it sounds like the right thing and cannot be confused with something else, you will be awarded the mark.

Read each question in detail – notice that part (c) has the important phrase 'shown on the diagram' which limits the acceptable answer. There are several features which help the small intestine to function, such as a rich blood supply and thin walls but only its length is shown on the diagram. Other similar phrases are often used to limit your possible answers, for example, 'not previously mentioned', or 'apart from ...' – look out for them!

What you should know at Credit level...

The important facts about proteins, carbohydrates and fats are shown in the table below. Digestive juices act on the large insoluble food molecules in food and convert them to smaller molecules which are soluble. This allows them to be absorbed into the blood and to be transported around the body.

The sites of production of the main digestive juices in a mammal are salivary glands, stomach, pancreas, liver and small intestine. These digestive juices contain enzymes, with the exception of **bile** from the liver which helps in the digestion of fats but does not contain the enzymes involved.

Type of Food	Protein	Carbohydrate	Fat
Use in the body	Growth and repair	Source of energy	Rich source of energy, also stores energy
Elements in the molecule	carbon hydrogen oxygen nitrogen	carbon hydrogen oxygen	carbon hydrogen oxygen
Enzyme used in digestion	proteases	amylase	lipase
Example of the enzyme	pepsin made in stomach	amylase made in salivary glands	lipase made in the pancreas
Product(s) of digestion	amino acids	simple sugars	fatty acids and glycerol

continued

What you should know at Credit level – continued

The structure of the villi in the small intestine is shown in the diagram.

The walls of the digestive system contain layers of circular muscle. Food is moved along by well coordinated contractions of these muscles. The muscles in front of and around the food relax and those behind it contract. This process is called **peristalsis**.

In a similar way, the walls of the stomach carry out powerful waves of contraction. This mixes the digestive juices thoroughly with the food. This means that each particle of food is likely to be fully exposed to the action of the digestive juice which greatly helps the chemical breakdown of food.

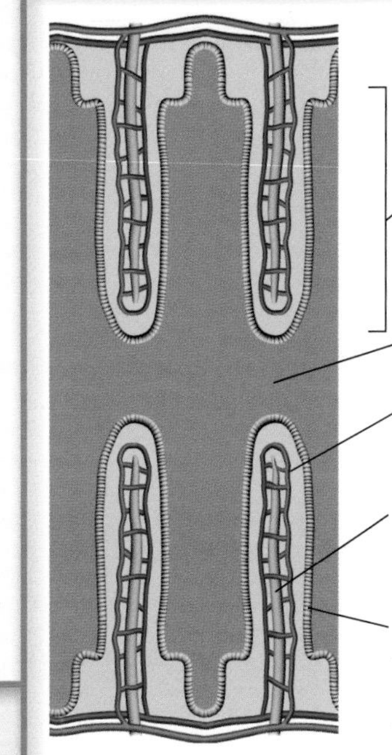

Villus – these tiny projections greatly increase the surface area for absorption of digested food by diffusion into the capillaries and lacteals.

Inside the small intestine – partially digested food mixed with enzymes. Soluble molecules are able to diffuse through the thin walls into the villi.

Blood capillaries – dissolved products of digestion diffuse into the blood and are transported throughout the body for use by the cells

Lacteal – connected to the lymphatic system. Collects the products of fat digestion for transport round the body.

Lining of small intestine – only one cell thick to allow dissolved products of digestion to diffuse from the intestine into the blood or lymph.

Credit question 1

The diagram shows part of the human digestive system.

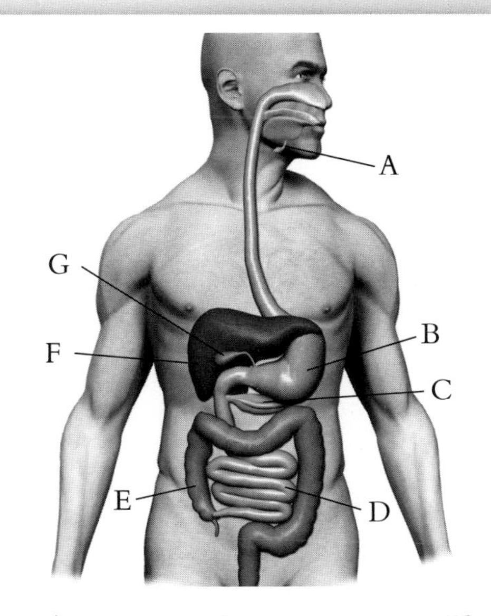

Use letters from the diagram to identify where the named digestive juices are produced. (2)

Credit question 1 – Answer

Digestive juice	*Letter*
pancreatic juice	C
saliva	A
gastric juice	B
bile	F

This question, and others, shows you that similar diagrams can crop up quite often. Bear in mind that the questions associated with them will almost always be different, and the labels may be in different places. So don't allow yourself to be casual about reading the question just because you recognise the diagram.

Credit question 2

A villus from the small intestine is shown below.

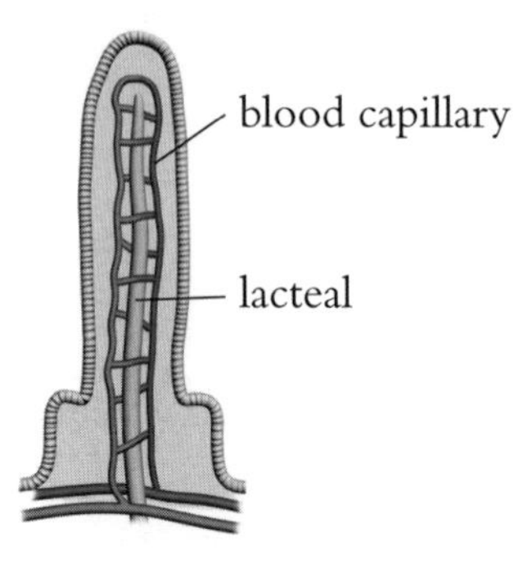

Describe two ways in which a villus is adapted for its role in the absorption and transport of food. (2)

Credit question 2 – Answer

Villi have a large surface area, thin walls, a rich blood supply, a lacteal.
(any two)

Credit question 3

The diagram below shows a stage in the movement of food along the intestine due to peristalsis.

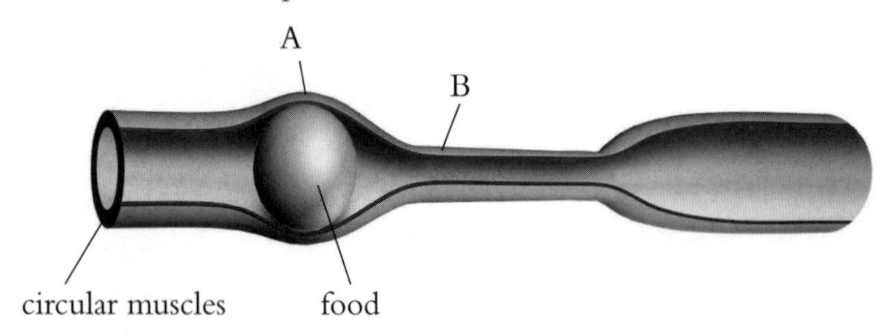

(i) Explain the mechanism of peristalsis by describing the muscular activity at points A and B. (1)
(ii) Which way does the food move in the diagram above? (1)

Credit question 3 – Answer

(i) Point A: *relaxed*
Point B: *contracted*
(ii) *to the left*

Credit question 4

Decide if each of the following statements about food substances and digestion is TRUE or FALSE, and tick the appropriate box. If the statement is FALSE then write the correct word(s) in the correction box, to replace the word(s) *underlined* in the statement. (3)

Statement	*True*	*False*	*Correction*
The elements present in fat are carbon, hydrogen and *nitrogen*.		✓	oxygen
Digestion involves the breakdown of *insoluble* food substances.	✓		
Carbohydrates are built up from *amino acid* molecules.		✓	glucose

*The true, false correction type of question is often used. You must be aware that you are only allowed to correct the **underlined** word or phrase. Lots of candidates suggest changes to the whole statement – and they lose the marks!*

Credit question 5

The table below contains the names of digestive enzymes, substrates and breakdown products.

Enzyme	*Substrate*	*Product (s)*
A pepsin	B fats	C glycerol and fatty acids
D lipase	E protein	F glucose
G amylase	H starch	I amino acids

Use the letters from the table to answer the following questions.

(i) Which enzyme is a protease? (1)

(ii) Identify the substrate and product(s) of the enzyme lipase. (1)

Credit question 5 – Answer

(i) A

(ii) Substrate: B

Product(s): C

Another example of questions which are easy marks for anyone who has revised the topic.

Reproduction

What you should know at General level...

Sexual reproduction involves:
- production of sex cells (**gametes**). **Sperm** are produced in **testes**; **eggs** are produced in **ovaries**.
- **fertilisation** – the fusion of the nuclei from one egg and one sperm.
- development of the fertilised egg(s) into young animals.

Fertilisation is the fusion of the nucleus from one sperm with the nucleus of an egg to make a fertilised egg cell. This is the first cell of the new animal and it has a complete set of genetic information from each parent.

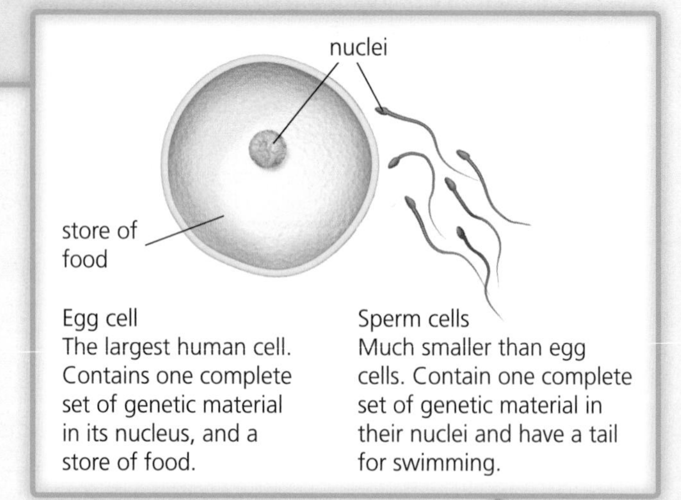

In most fish sperm and eggs are deposited in the water and so fertilisation takes place outside the body of the parents. This is called **external fertilisation**. Fish eggs have a flexible covering and contain a supply of food called **yolk**. When they hatch, the young feed on the remains of the yolk for a few days and are then able to maintain themselves. In most species of fish the young have no parental care.

In mammals, the sperm are deposited inside the female and so fertilisation takes place inside her body. This is called **internal fertilisation**. Internal fertilisation is necessary for land dwellers since sperm need fluid to swim to the egg. The fertilised egg moves down the **oviduct** to the **uterus** and burrows into the uterine wall. This is called implantation. The fertilised egg develops into an **embryo** and then into a **fetus**. The fetus develops in a bag of watery fluid called the amniotic sac and gets its food from the mother's blood system. The young leave the uterus at birth and are dependent upon the adult for care and protection.

The names and functions of human reproductive organs are shown on the diagrams.

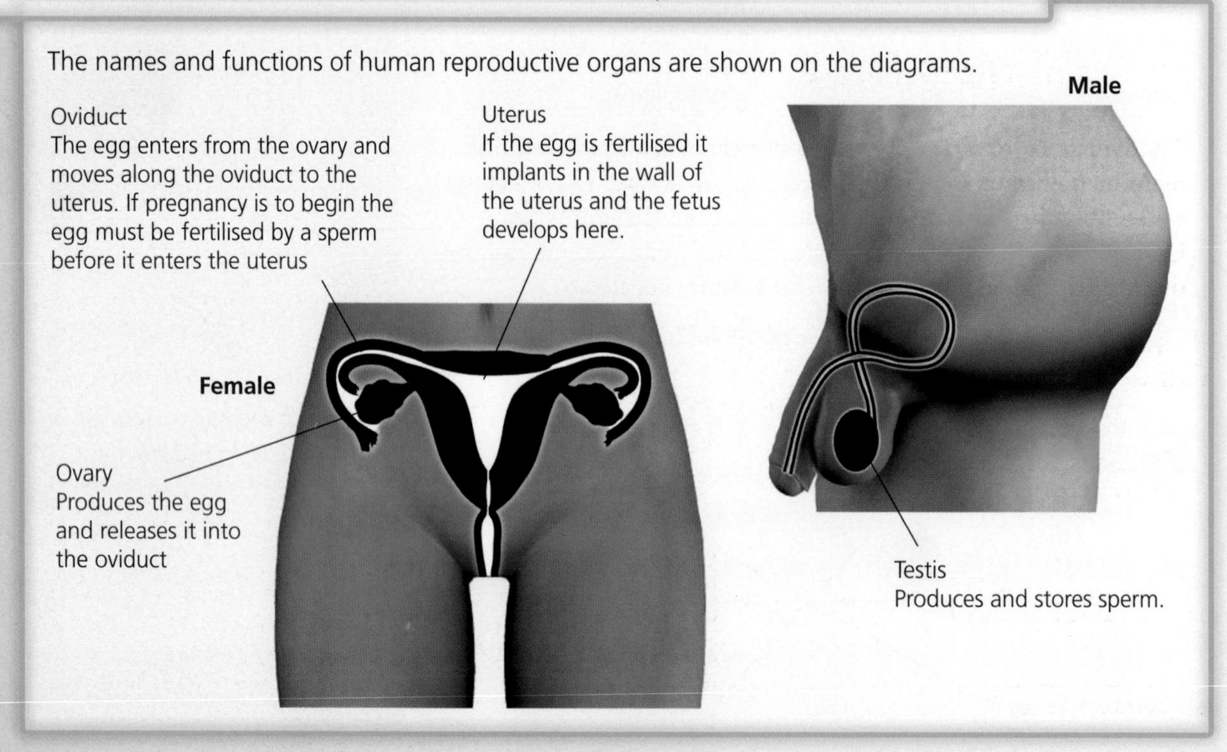

General question 1

The diagrams below show male and female gametes from a number of different animals.

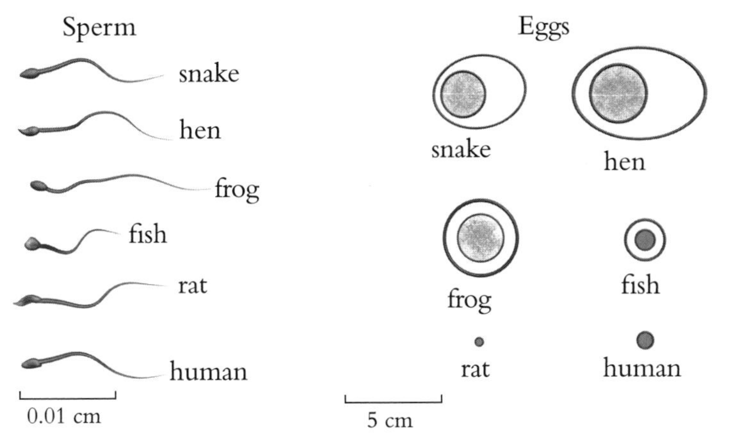

(a) State one structure shown on the diagrams which is common to all of the sperm cells. Explain the function of this structure. (2)
(b) Explain why the fish egg is larger than the human egg. (1)

General question 1 – Answer

(a) Structure: *tail* (1 mark)
Function: *allows the sperm to swim towards the egg* (1 mark)
(b) *Fish embryos rely on the yolk of the egg for food. Human embryos obtain their food from the mother's blood.*

The phrase 'shown on the diagram' is important. An answer about a feature that is not shown will not get the mark – even if it is true.

General question 2

The information in the boxes below refers to stages of reproduction in mammals and to parts of the female reproductive system.
Use arrows to join the boxes on the left with those on the right so that the part where each stage occurs is shown correctly. (2)

General question 2 – Answer

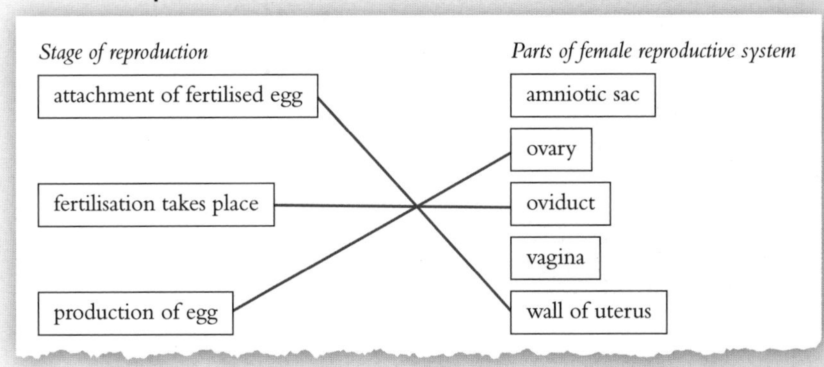

General question 3

The diagram shows an egg cell about to be fertilised by a sperm.

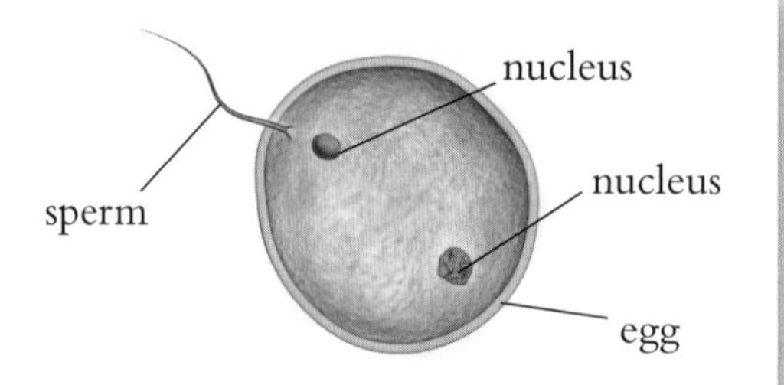

Describe what would happen next, to complete fertilisation. (1)

General question 3 – Answer

The nucleus of the sperm fuses with the nucleus of the egg.

General question 4

Where are human sperm cells produced? (1)

General question 4 – Answer

In the testes

General question 5

Name the organ in which a human fetus completes its development. (1)

General question 5 – Answer

The uterus

Once again, notice the big rewards in terms of marks for learning the basic facts. Try to use proper biological terms if possible. The main reason is that they give the exact answer rather than an everyday word which may mean different things to different people. Obviously this is especially important in the area of reproduction, when some of the words and names of parts you may hear will certainly not gain marks in the national examination!

General question 6

The diagram represents the reproductive system of a human female.

(a) Name the parts labelled on the diagram. (2)

(b) In the table below, match each letter from the diagram to its correct function. (2)

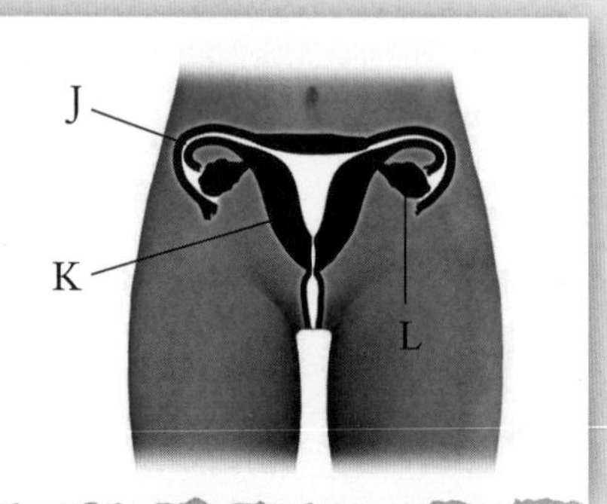

General question 6 – Answer

(a) J *oviduct*
K *uterus*
L *ovary*

(b)

Function	*Letter*
Eggs produced	L
Fertilisation takes place	J
Fertilised egg becomes attached	K

General question 7

Tick (✓) boxes in the table to indicate whether each of the following statements is true for eggs, sperm, or both. (2)

General question 7 – Answer

Statement	*Eggs*	*Sperm*
Contain a food store for developing fetus	✓	
Swim using a tail		✓
Produced in testes		✓
In most fish, are deposited into the water	✓	✓
Are gametes	✓	✓

The tick boxes show a very important point. Sometimes you need a tick in more than one box in each row, and sometimes not. The first three rows could easily lull you into thinking that each statement applied to either eggs or sperm, but clearly both are deposited in water, and both are gametes. There is a hint at the end of the question to alert you – but only if you are reading and thinking thoroughly.
Notice that the same answers come up over and over again in different questions. The way the question is worded, or how the answer is laid out (lines for answers, complete a table, tick boxes and so on) can change, but the basic facts stay the same. Don't let a new way of asking a question worry you.

What you should know at Credit level...

The number of eggs which any animal needs to produce for successful reproduction depends on the chance of fertilisation and the degree of parental care. Internal fertilisation increases the chances of fertilisation since the sex cells are closer together when released, and this means that fewer eggs need to be produced. Because fish usually have a lower chance of fertilisation and give little or no protection they need to produce far more eggs than mammals to ensure the survival of their species.

continued

What you should know at Credit level – continued

Embryo mammals develop a **placenta**, to which they are connected by the **umbilical cord**. The placenta allows food and oxygen to pass from the mother's circulation to the embryo, and waste materials and carbon dioxide to pass in the other direction It is possible for many harmful substances such as drugs to pass from the mother to the embryo through the placenta if she takes them into her body.

Umbilical cord
Contains blood vessels which obtain food and oxygen from the placenta and return waste materials and carbon dioxide to the placenta.

Placenta
A network of blood vessels from both the fetus and the mother. The blood does not mix, but the vessels are closely entwined allowing food and oxygen to diffuse from the mother's blood to the fetus and carbon dioxide and waste to diffuse from the blood of the fetus to the mother's.

Wall of the uterus
This is made of muscle and expands as the fetus grows. During birth it contracts powerfully to push the baby through the birth canal.

Fetus
Started as a single fertilised egg and obtains its food from the mother's blood,

Amniotic sac
Fluid filled membrane which protects the fetus until birth.

Credit question 1

(a) The diagram below shows the uterus of a pregnant human female. Draw a line from the word placenta to show its position on the diagram. (1)

Credit question 1 – Answer

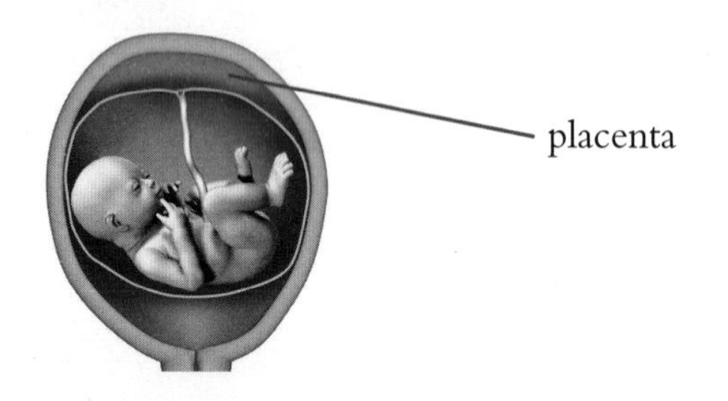

Make sure that the ends of the line are clear and touch the correct area, not just pointing to it. It is a good idea to use a ruler. A significant proportion of candidates miss out questions which ask them to add a label or a line to a diagram. The reason seems to be that the lack of a space for an answer makes them fail to notice the question at all. This means the loss of easy marks simply due to lack of careful reading.

Credit question 2

Which of the following statements are correct? (1)

1 The placenta is a large disc of tissue attached to the wall of the uterus.

2 The placenta has a rich supply of vessels connected to the umbilical cord.

3 The blood vessels of the embryo connect with those of the mother in the placenta.

Credit question 2 – Answer

1 and 2

Statement 3 is wrong because the blood of the embryo and mother never mix. The vessels are very close to each other but do not connect.

Credit question 3

From the data in the table, explain the relationship between the number of eggs or young produced and the chance of survival to maturity. (1)

	Usual number of eggs or young produced at one time	*Length of time of parental care*
Cod	5,000,000	None
Stickleback	300	Weeks
Dog	4	Months
Human	1	Years

Credit question 3 – Answer

Increased parental care increases the chance of survival of the eggs and young so fewer eggs or young need to be produced.

Make sure you understand cause and effect correctly. For example, the degree of parental care determines how many eggs need be produced, not the other way round.

Water and waste

What you should know at General level...

In all mammals the water taken in must equal the water lost.

Water is gained:
- from drinking
- from water contained in food
- from chemical reactions in the body.

Water is lost:
- as sweat
- as water vapour when breathing out
- in faeces
- in urine.

The **kidneys** are the main organs which a mammal uses to adjust its water content.

Blood is brought to the kidney in the **renal artery**. The kidneys remove urea, glucose, salts and water from the blood by **filtration**. Useful materials such as the glucose and some of the water are returned to the blood by **reabsorption**. The blood then returns to the circulation through the **renal vein**.

The materials not reabsorbed form **urine**, which is taken from the kidneys to the **bladder** by the **ureters**. The main waste product removed in the urine is **urea**.

continued

What you should know at General level – continued

The diagram shows the names and functions of the structures involved.

Since the kidney is responsible for the removal of wastes from the blood, any damage, either from accidents or disease, can lead to a build up of poisonous wastes in the body. We can survive on one kidney, but total kidney failure would be fatal if not treated by dialysis on a kidney machine, or by a kidney transplant.

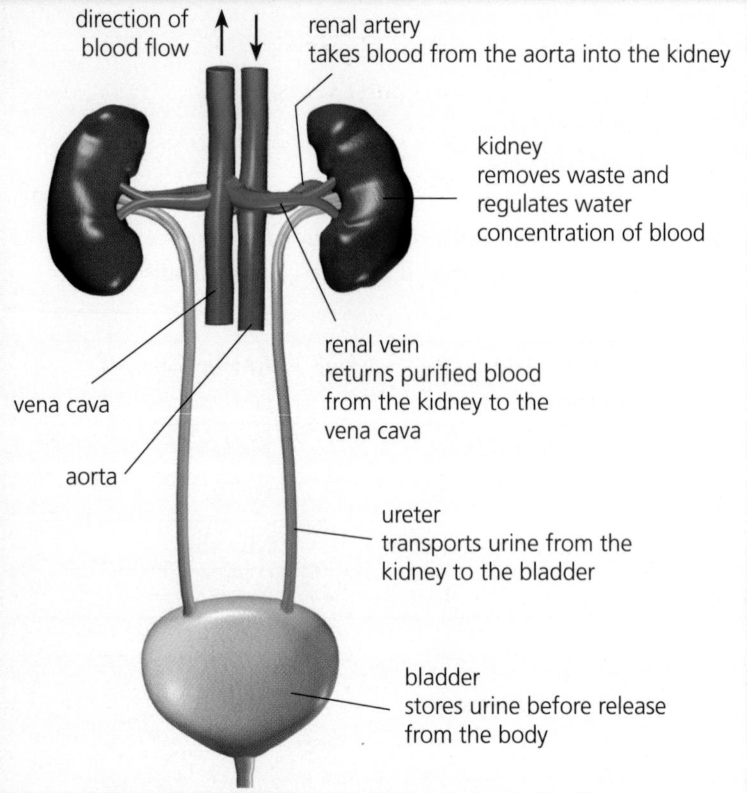

General question 1

The diagram below shows a healthy human kidney.

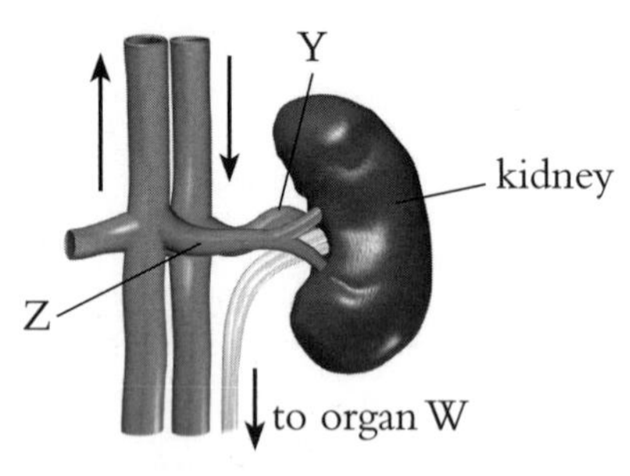

Name the structures labelled X, Y and Z and organ W. (2)

General question 1 – Answer

X *ureter*
Y *renal artery*
Z *renal vein*
W *bladder*

*Take care to spell **ureter** correctly. It can easily be confused with the word **urethra** so if you spell it wrong, you will lose a mark.*

General question 2

Complete the following sentences using the words from the list.

List vein artery glucose protein excreted reabsorbed (2)

General question 2 – Answer

Blood is taken to the kidneys in the renal *artery* .
In the kidney *glucose* is filtered out of the blood and then *reabsorbed* .

Look out for

Arteries carry blood away from the heart, (remember **A**rteries **A**lways **A**way) so the blood vessels entering all the main organs are arteries.

All dissolved substances in the blood are filtered out, but glucose is so valuable that it is all reabsorbed again.

General question 3

Complete the table to show the daily gains and losses of water by a small mammal. (2)

General question 3 – Answer

Water gain (cm^3)		*Water loss* (cm^3)	
food	170	*urine*	300
drink	*260*	faeces	*22*
chemical reactions	70	sweat	100
Total	500	breath	78
		Total	500

This is a common type of question. You need to know the routes by which water is gained and lost as well as being able to do the simple calculations. There is only one gain and one loss missing, so there is only one answer in each case.

- *The total gain is 500, so drink must be:*
  500 – 170 – 70 = 260
- *Water loss is 500 so faeces must be:*
  500 – 300 – 100 – 78 = 22

What you should know at Credit level...

The kidney adjusts the volume of urine produced under the control of the hormone **ADH**.

The brain monitors the water concentration of the blood.

If the water concentration of the blood falls:

- more ADH is produced
- this makes the kidney reabsorb more water
- a smaller volume of concentrated urine is produced
- this reduces water loss.

If the water concentration of the blood rises:

- less ADH is produced
- this makes the kidney reabsorb less water
- a larger volume of dilute urine is produced
- this increases water loss.

Urea is a waste product produced when surplus amino acids are broken down in the liver. Urea is then transported in the blood to the kidneys where it is filtered out and removed.

Each kidney has many millions of **nephrons** which are the microscopic structures which produce urine. Blood is filtered in the **glomerulus**. The filtrate is collected by the **Bowman's capsule** and enters the **tubules** where useful substances such as glucose, some salts and water are reabsorbed into the blood.

Reabsorption is done by blood capillaries which are closely wrapped around the tubules. The waste, consisting of water, some salt and urea is called urine and it is collected by the **collecting ducts**, taken to the ureters and then to the bladder.

continued

What you should know at Credit level – continued

Kidney machines can keep patients alive until a transplant becomes available, but they have several disadvantages:

- the patient must have his or her blood connected to the machine for several hours every week.
- patients must follow a very rigid diet to avoid complications.
- machines only work for a limited time for each patient.

Kidney transplants can save a patient, and give him or her a relatively normal life. But kidney transplants also have disadvantages:

- any major surgery carries some risk.
- the kidney may be rejected by the body of the patient.

The diagram shows the detailed structure of a kidney nephron and the functions of the different parts.

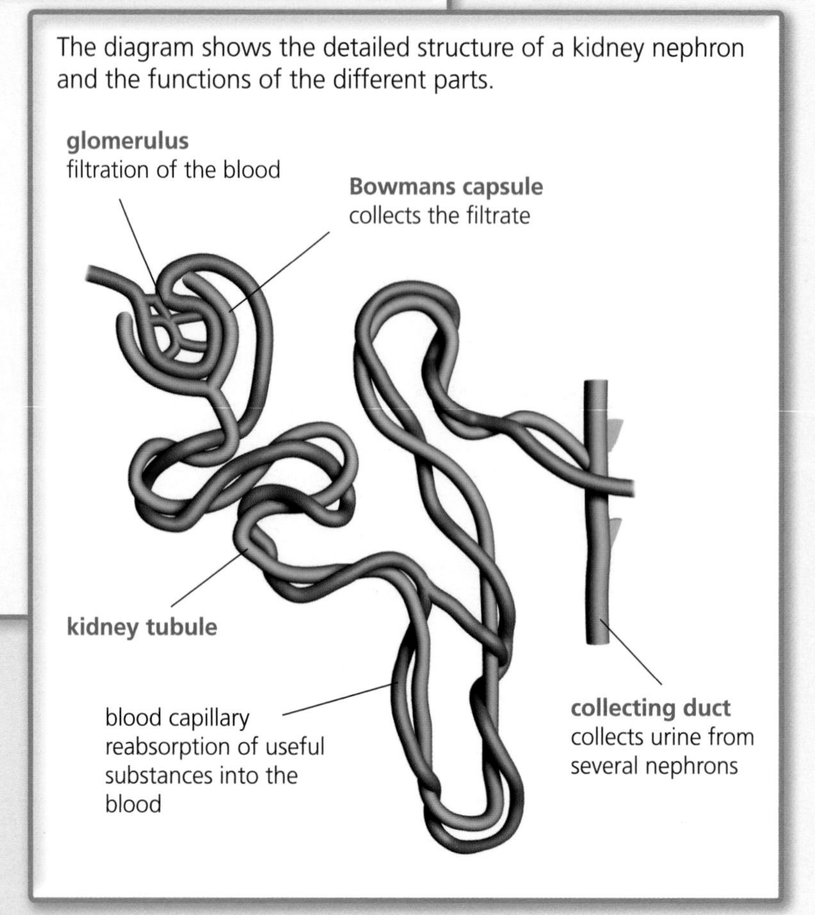

Credit question 1

The statements below refer to factors which affect the level of the hormone ADH in the blood.

1 Drinking a large volume of water
2 A low water concentration in the blood
3 Losing sweat when running
4 A high water concentration in the blood

Which two factors would bring about a decrease in the level of ADH in the blood? (1)

Credit question 1 – Answer

....1.... and ....4....

ADH is the signal to conserve water and is present in the blood when the water concentration of the blood is low. Drinking a large volume of water (statement **1***) would lead to a high water concentration in the blood (statement* **4***), both of which would cause a decrease in the level of ADH in the blood*

Credit question 2

The diagram represents a kidney nephron.

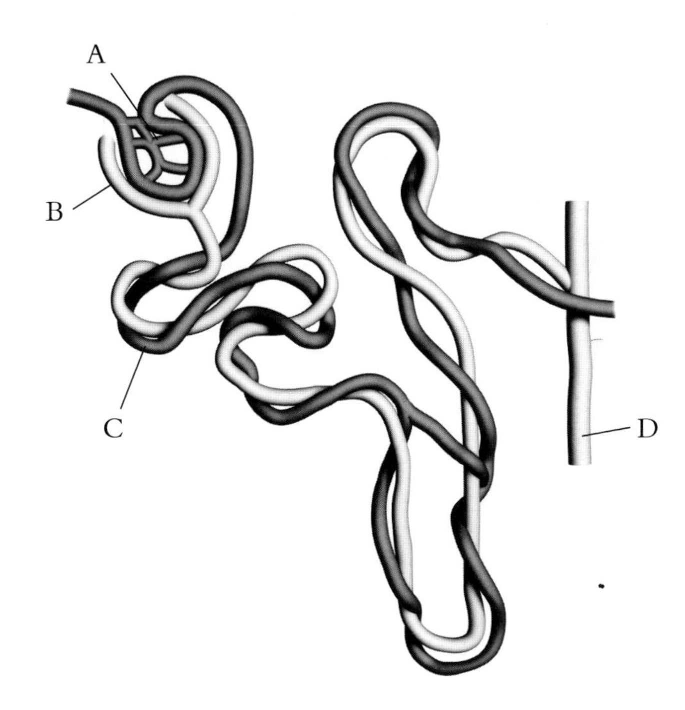

Complete the table to show the name and function of the labelled parts of the nephron. (2)

Credit question 2 – Answer

Letter	*Name*	*Function*
A	glomerulus	*filters blood*
B	*Bowman's capsule*	collection of filtrate
C	*blood capillary*	reabsorption
D	collecting duct	*collects urine*

Again, these are easy marks for simple recall of factual material. Learning the facts always pays off!

Responding to the environment

What you should know at General level...

All animals have to respond to changes in their environment if they are to survive. Light, humidity and chemicals are three environmental factors which affect behaviour. A factor which produces a **response** is called a **stimulus**.

For example:

- maggots or woodlice move away from the light
- woodlice move towards moisture (humidity)
- many animals move towards (or away from) the smell of different chemicals.

Rhythmical behaviour in animals results from regular events in the environment, such as tides, day/night, or seasons. The factor which causes and controls rhythmical behaviour is called the **trigger stimulus**. For example:

Rhythmical behaviour	*Trigger stimulus*
Annual migrations by birds	Changes in day length
Feeding in cockles, mussels on the sea shore	Rise and fall of the tides
Hunting at night by owls	Darkness

In every case the response of an animal to an environmental stimulus is important for its survival.

General question 1

The apparatus shown was set up to investigate the behaviour of woodlice.

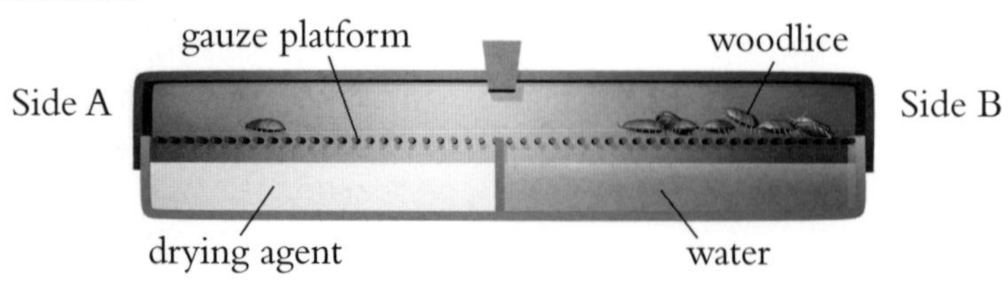

(a) At the start of the investigation 20 woodlice were placed in the centre of the chamber. After 10 minutes there were 2 on side A and 18 on side B.

(i) What environmental factor was being investigated? (1)

(ii) Describe the response of the woodlice in the investigation. (1)

(b) Other than the factor above, name one other environmental factor, and describe the response of a named animal to that factor. (2)

General question 1 – Answer

(a) (i) *moisture or humidity*
(ii) *the woodlice moved towards the area of higher moisture*

(b) Factor: *light*
Named animal: *maggots*
Response: *they move away from light*

In part (b) you are asked for one example, so it doesn't matter which environmental factor you pick (apart from moisture which is already mentioned). The marks are for naming an animal and giving an appropriate response it makes to the factor you chose.

*When you describe the response of an animal you must only say what it actually does. DO NOT say what it thinks. For example the answer 'woodlice prefer damp conditions' would NEVER get a mark. Words such as **prefer**, **like**, **hate** suggest human emotions which we cannot assume that any animal has. They will automatically lose you the mark.*

What you should know at Credit level...

It is important to be able to explain how an animal's response helps it to survive. For example:

- by moving away from light the maggots and woodlice become less visible to predators.
- by moving towards moisture the woodlice will keep their gills working efficiently.
- by responding to smells animals may find food, avoid danger, or find potential mates.

In the same way it is important to be able to explain how rhythmical behaviours are significant in the lives of the animals concerned. For example:

- migration ensures that breeding happens in the best conditions even if the rest of the year is better spent in a different habitat.
- many animals can only feed safely when covered by the tide and remain inactive and hidden from land-based predators when the tide is out.
- the owl hunts prey which is nocturnal and the prey has little chance of seeing it coming.

Credit question 1

Complete the following table by suggesting a way in which each of the behaviours described might contribute to the survival of the animal. (2)

Credit question 1 – Answer

Animal	*Behaviour*	*How survival chances are improved*
Housefly	moves toward the smell of decaying meat	*increases the chances of finding food*
Young deer	lies completely still when it hears movement nearby	*reduces the chance of being found by a predator*

You are not expected to have learned the answer in this case – you need to use your understanding of the basic principles to deduce a possible answer.

Investigating living cells

What you should know at **General** level...

The bodies of living organisms are built up of small structures called **cells**. Some organisms consist of just a single cell. Others can contain many millions of cells. Cells are very small and can only be seen using a microscope.

Cells are usually transparent and difficult to see even with a microscope. Chemicals called **stains** are often added to cells to make parts of the cells easier to see with a microscope.

The cells of all organisms have similar structures but there are some differences between plant cells and animal cells. The diagram shows the main structures in typical plant and animal cells.

Diagram of a plant cell	*Name and function of cell structures*	*Diagram of an animal cell*
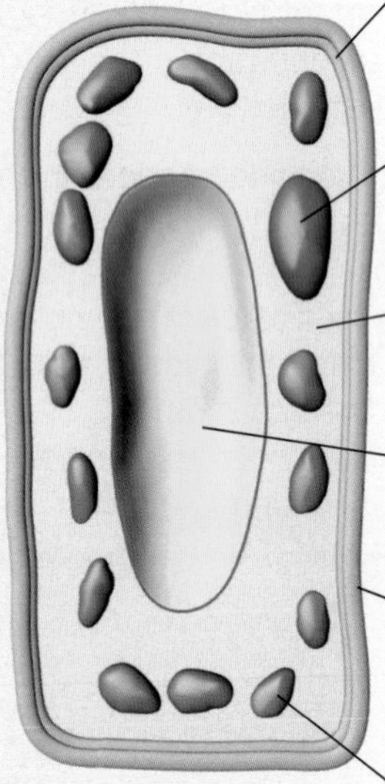	**Cell membrane** – this surrounds the living cell. It controls what goes in and what comes out of the cell	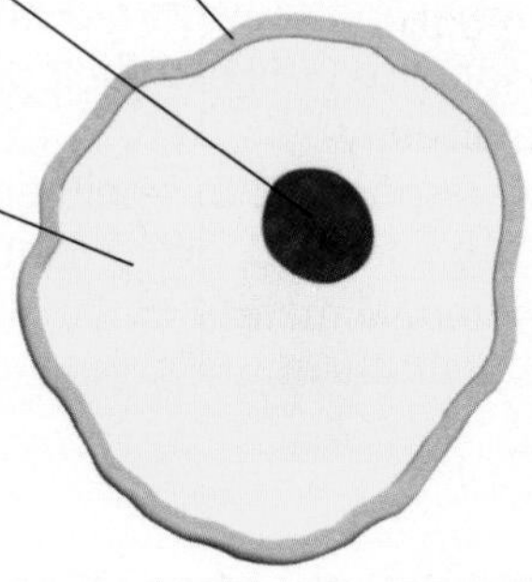
	Nucleus – this contains the genetic information that controls the development and activities of the cell.	
	Cytoplasm – this is a jelly-like material. It is where the chemical reactions of the cell take place.	
	Vacuole – this is a space filled with cell sap and it keeps the cell firm. It is only present in plant cells.	
	Cell wall – this is a strong outer layer surrounding plant cells. It gives support to the plant tissues.	
	Chloroplast – these contain the chemical chlorophyll and they are the sites of photosynthesis. They are only found in the green parts of plants.	

General question 1

The diagram shows some of the structures found in a typical plant cell.

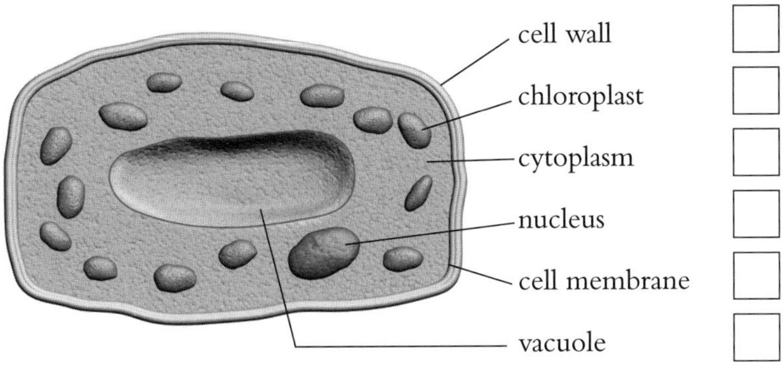

cell wall	☐
chloroplast	☐
cytoplasm	☐
nucleus	☐
cell membrane	☐
vacuole	☐

Tick the boxes to show the structures that are also found in a typical animal cell. (2)

General question 1 – Answer

cell wall	☐
chloroplast	☐
cytoplasm	✔
nucleus	✔
cell membrane	✔
vacuole	☐

Look out for

Remember there are only three parts to a typical animal cell. These same parts are in plant cells, but plant cells have additional structures.

General question 2

Why are cells often stained before being used under a microscope? (1)

General question 2 – Answer

To make the parts of the cell easier to see.

If you are asked for the general name of substances used to do this, 'stain' is the only acceptable answer. If you answer 'dye' or give the name of an actual stain – you will lose the mark.

Investigating diffusion

What you should know at General level...

The molecules of liquids and gases are continually moving. This causes them to spread and move from an area of high concentration to areas of lower concentrations. This movement is called **diffusion**.

Many molecules are small enough to diffuse through cell membranes but others are too large. In this way, the cell membrane controls the movement of substances into and out of a cell. Substances that can enter or leave cells by diffusion include water, oxygen, carbon dioxide and dissolved foods such as glucose.

The diffusion of water through a cell membrane is called **osmosis**.

General question 1

The concentration of some substances inside and outside three cells is shown in the diagrams below.

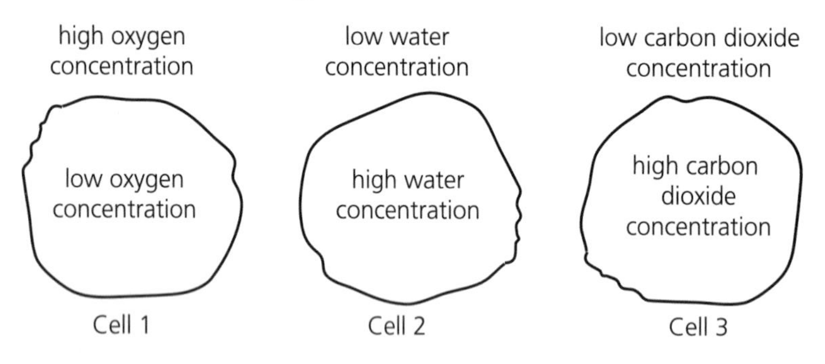

Use the cell numbers from the diagrams to identify the cells in which the following would occur. (2)

General question 1 – Answer

(i) Diffusion **into** the cell of the substance shown: *Cell 1* (1 mark)

(ii) Osmosis: *Cell 2* (1 mark)

(i) Substances always diffuse from high to low concentrations. Cell 1 is the only one with the high concentration on the outside.

(ii) Osmosis only ever involves the movement of water. Cell 2 is the only one involving water movement.

What you should know at Credit level...

Cells can gain useful substances and get rid of waste substances by diffusion. For example, the glucose and oxygen needed for respiration enter cells by diffusion. The carbon dioxide produced during respiration leaves cells by diffusion.

Osmosis refers to the diffusion of water into or out of cells. Cell membranes are **selectively permeable**. This means that they allow small molecules to pass through them but large molecules are prevented from passing through.

continued

What you should know at Credit level – continued

If the water concentration is different on each side of a cell membrane, we say that a **concentration gradient** exists. Water passes down the concentration gradient. In other words, it passes from the high to the low water concentration, through the cell membrane. This is osmosis.

The diagrams show what happens to cells placed in high or low water concentrations.

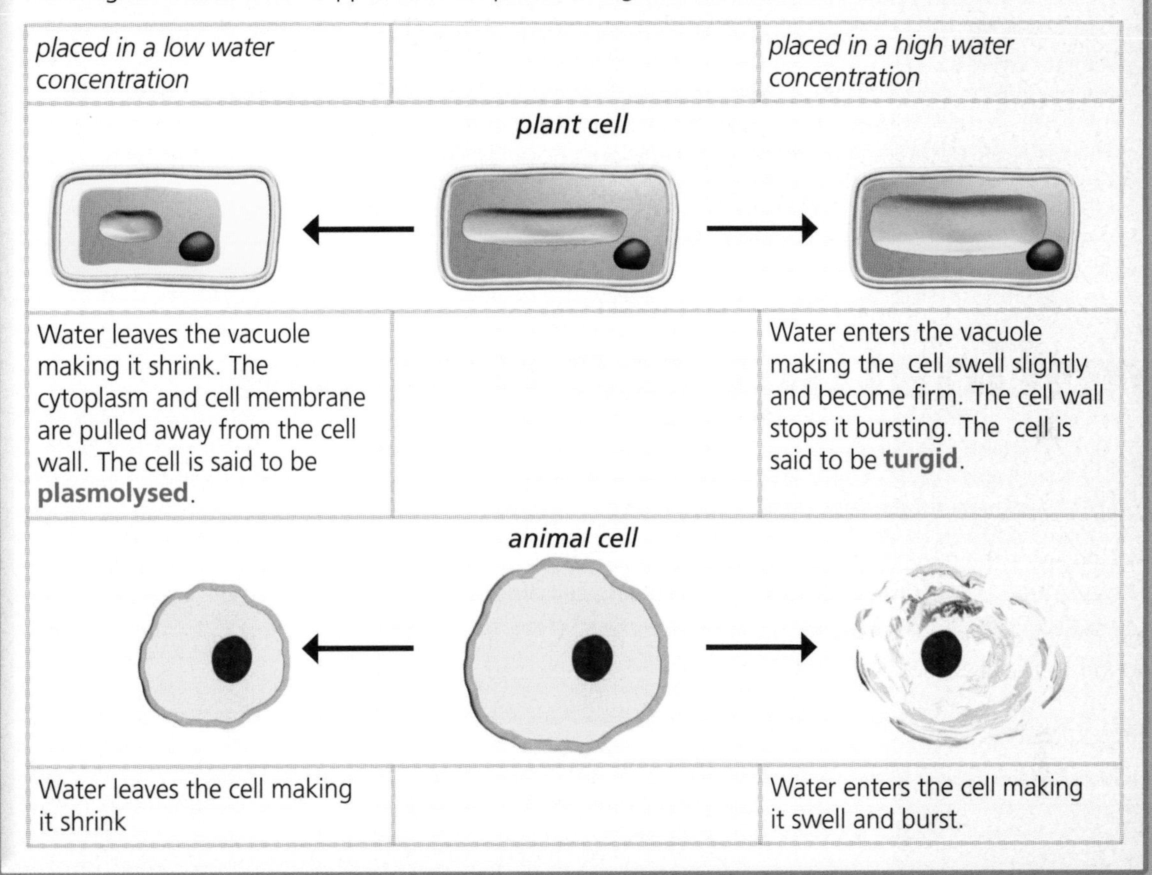

placed in a low water concentration		*placed in a high water concentration*
	plant cell	
Water leaves the vacuole making it shrink. The cytoplasm and cell membrane are pulled away from the cell wall. The cell is said to be **plasmolysed**.		Water enters the vacuole making the cell swell slightly and become firm. The cell wall stops it bursting. The cell is said to be **turgid**.
	animal cell	
Water leaves the cell making it shrink		Water enters the cell making it swell and burst.

Credit question 1

A flower petal was examined under the microscope and then placed in a concentrated salt solution for 30 minutes. It was then re-examined under the microscope.
The diagrams show a cell from the petal before and after being in the solution.

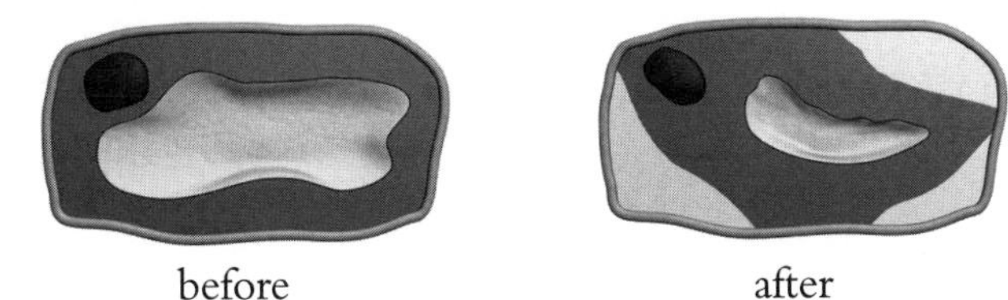

(i) The movement of water caused the change in the appearance of the cell. What name is given to this movement of water? (1)

(ii) In terms of water concentration, explain **why** this movement of water took place. (1)

Credit question 1 – Answer

(i) *Osmosis*
(ii) *The water concentration was higher inside the cell than outside. Water moved out of the cell from the higher water concentration inside.*

To answer this question you must say where the high and low water concentrations are. You must then say in which direction the water will move. Concentrated solutions have low water concentrations so there will be a higher water concentration inside the cell. The water moves by osmosis from inside the cell to the outside.

Investigating cell division

What you should know at General level...

Cell division increases the number of cells in an organism. This means that it is essential for growth. The cell nucleus controls all cell activities, including cell division.

The division of the cell nucleus is called **mitosis**. Mitosis makes sure that the number and type of **chromosomes** present in the two daughter cells is exactly the same as those of the parent cell. This means that the genetic information of both daughter cells will be exactly the same as that of the parent cell.

The stages in cell division are shown below.

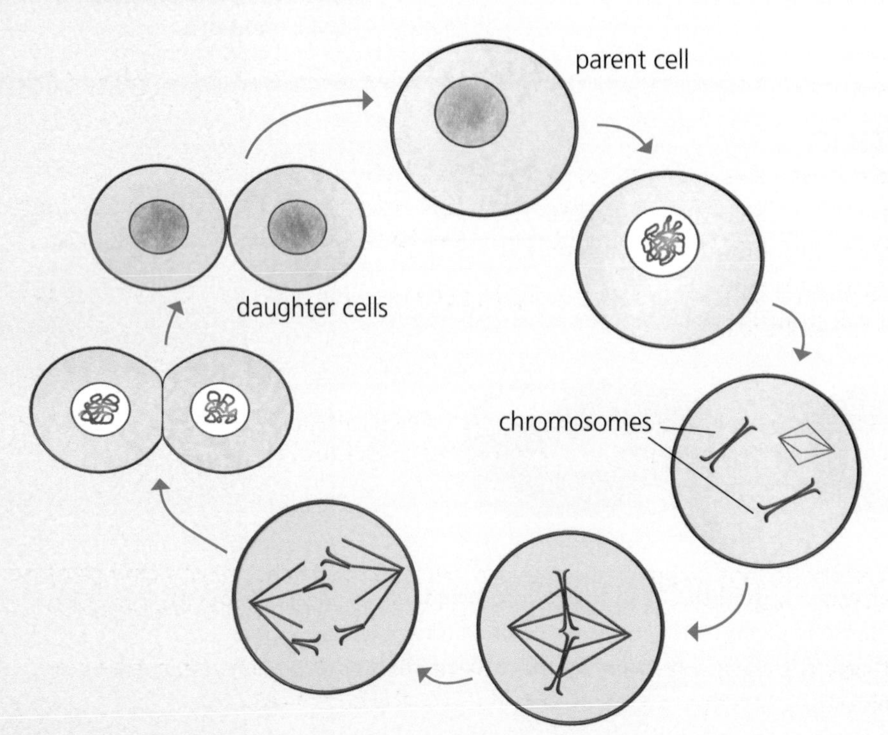

General question 1

The following diagrams show four stages of mitotic cell division but not in the correct order.

A 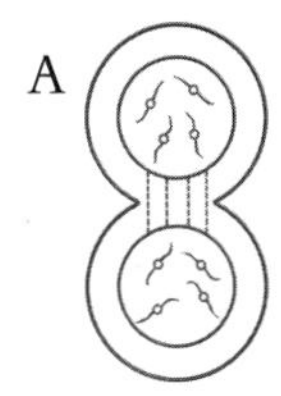 B 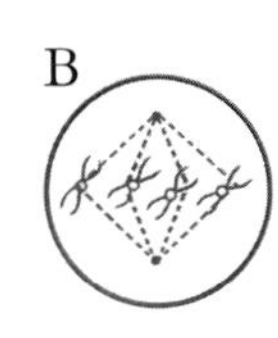 C 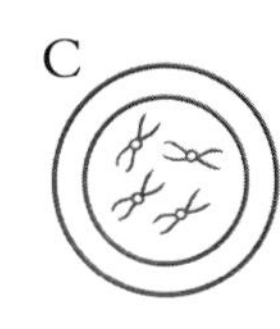 D 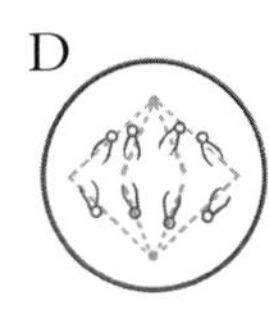

Arrange the letters from the diagrams to put the stages into the correct order. The first stage has been given. (1)

General question 1 – Answer

1st stage C
2nd stage B
3rd stage D
4th stage A

Diagrams of this process may differ in their appearance. The important point is that you know what is happening to the chromosomes in the cell and that you can follow the changes in their appearance.

What you should know at **Credit** level...

At each stage of cell division by mitosis the chromosomes go through a sequence of changes.

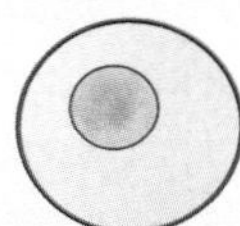

Before division begins, each chromosome in the nucleus is duplicated even though they cannot be seen.

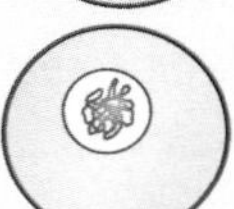

The chromosomes become visible as strands in the nucleus.

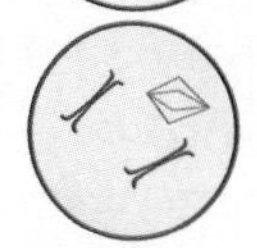

The nuclear membrane breaks down and each chromosome is visible as a pair of **chromatids**. A set of fibres, the spindle appears in the cell. This cell has two chromosomes but the cells of most organisms have more. Human cells have 46 chromosomes.

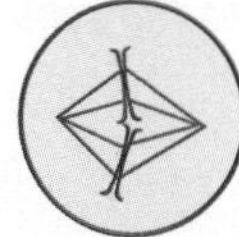

Each chromosome, or chromatid pair, becomes attached to a **spindle fibre** near the middle of the cell.

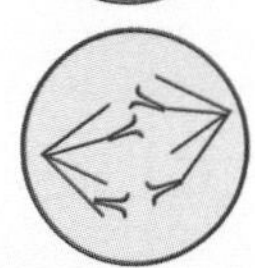

The spindle fibres contract and separate the chromatids in each pair. The chromatids from each pair are pulled to opposite ends of the cell.

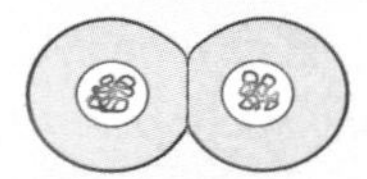

The chromatids begin to become less distinct and each becomes a new chromosome. New nuclear membranes begin to form and the cytoplasm begins to divide.

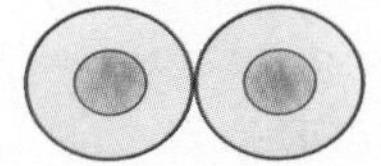

Division is complete. There are now two daughter cells each with a full set of chromosomes that are identical to those of the parent cell.

continued

What you should know at Credit level – continued

Mitosis is the same in animal and plant cells. The only difference is the way in which the final separation into two cells takes place. In animal cells the cell membrane becomes pinched between the two nuclei and divides the cytoplasm into two. In plant cells a new cell wall is formed to divide the cell into two.

It is important that the **chromosome complement** (the number and type of chromosomes) of the daughter cells is the same as that of the parent cell. This is because the chromosomes contain the genetic information which controls the development and activities of the cell. Any change or loss of chromosomes would mean that the cell would not function properly.

Credit question 1

The diagram below contains some of the stages of cell division by mitosis.
Describe stage **2** and **5** in the spaces provided. (2)

Credit question 1 – Answer

Stage 1
Chromosomes become visible as pairs of chromatids.

Stage 2
Chromatid pairs **or** *chromosomes attach to spindle fibres*
or
Chromatid pairs **or** *chromosomes gather at cell equator*

Stage 3
The spindle fibres contract, pulling the chromatids of each chromosome to opposite poles of the cell.

Stage 4
A nuclear membrane forms around each nucleus.

Stage 5
The cell divides into two daughter cells.

There is often more than one answer to this type of question. As long as your answer refers to a particular event that takes place between the stage immediately before and the stage immediately after the one you are asked about, you will get the mark.
*In this case, Stage 2 could be any relevant point that takes place between the appearance of the chromosomes and the separation of the chromatid pairs. You must take care **not** to say that chromosome pairs separate. The chromosomes are not in pairs during mitosis but each chromosome is made up of a pair of chromatids.*
Stage 5 must refer to the final separation into two cells.

Credit question 2

Mitosis ensures that all daughter cells in a multicellular organism have the same number and type of chromosomes.
Explain why this is necessary. (1)

Credit question 2 – Answer

So that there is no loss of genetic information.
or *So that every cell has all the correct genetic information.*

This question is about individual cells – not whole organisms. Any answer that refers to possible abnormalities in an organism will lose the mark.

Investigating enzymes

What you should know at General level...

Some chemical reactions need **catalysts** to make them happen. A catalyst is a substance which speeds up a chemical reaction but which is left unaltered after the reaction.

All cell activities are the results of chemical reactions in the cell.

Enzymes are catalysts that speed up the chemical reactions of cells. Every reaction in a cell needs an enzyme to make it happen. One enzyme molecule can make the same chemical reaction happen many times.

All enzymes are made of protein. Different enzymes are made of different proteins.

Some enzymes cause the breakdown of substances into smaller molecules. Examples are the enzyme **catalase** and the digestive enzyme **pepsin**. Their actions can be represented as:

protein $\xrightarrow{\text{pepsin enzyme}}$ peptides (short chains of amino acids)

hydrogen peroxide $\xrightarrow{\text{catalase enzyme}}$ water + oxygen

Some enzymes cause the build up of substances from smaller molecules. An example is the enzyme **phosphorylase**. Its action can be represented as:

glucose –1 – phosphate $\xrightarrow{\text{phosphorylase enzyme}}$ starch

The activity of an enzyme is affected by temperature. At 0°C most enzymes will not work. As the temperature increases, enzyme activity increases until the temperature reaches a point where enzyme activity is at its maximum. As the temperature increases further, enzyme activity decreases rapidly and stops. This is because the enzymes have become **denatured** or permanently damaged by the heat. This effect can be shown as a graph.

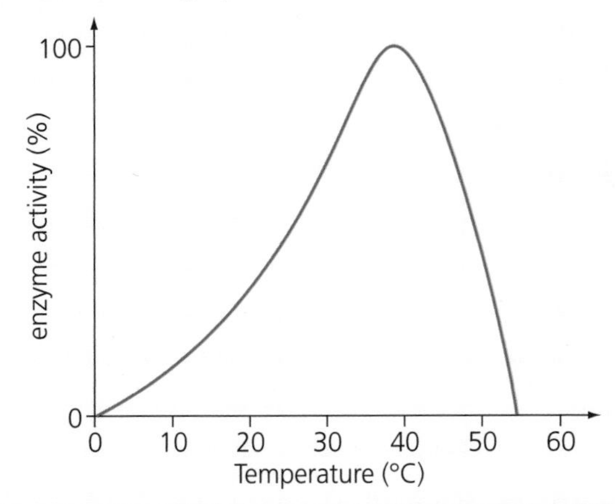

The activity of an enzyme is also affected by pH. There is a pH value for every enzyme at which it is most active. The enzyme will work within a range of pH values but will be inactive outside of this range. This effect is shown for the enzymes pepsin and catalase in the graph.

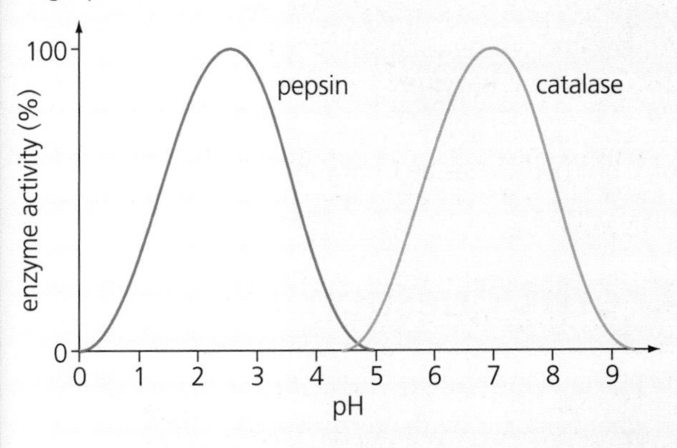

General question 1

Enzymes are biological catalysts.
Explain the meaning of the word *catalyst*. (1)

General question 1 – Answer

A chemical which speeds up a chemical reaction and which is unchanged at the end of the reaction.

There are two parts to this answer and both are needed to get the mark.

General question 2

Samples of trout eggs were kept at different temperatures.
The number of days it took the eggs to hatch is shown in the table

Temperature (°C)	5	10	15	20	25	30	35	40
Time to hatching (days)	62	42	24	10	4	6	18	did not hatch

Explain why the eggs kept at 40°C did not hatch. (1)

General question 2 – Answer

The temperature was too high and the eggs died. **or**
The enzymes in the eggs were denatured by the high temperature.

Remember that every biological process depends on enzymes and so the effect of increasing temperatures will be the same for them all. Increasing temperatures will increase the rate at which the process takes place. At a particular temperature the enzymes will be most active and further increases in temperature will cause the process to slow and stop as the enzymes are denatured.

What you should know at Credit level...

In enzyme-controlled reactions, the substances used up are called the **substrates**. The substances that are produced are called the **products**. Each enzyme will work for only one type of chemical reaction. It will act on only one substrate. The enzyme is said to be **specific** for that reaction or for that substrate.

The conditions in which an enzyme is most active are called the **optimum** conditions.

An enzyme will have an optimum temperature for its activity. This is normally about 38°C for most enzymes.

An enzyme will also have an optimum pH for its activity. Catalase has an optimum pH of 7 but pepsin has an optimum pH of 2.5.

Credit question 1

The enzyme maltase is specific for the sugar maltose. Explain what is meant by the term specific. (1)

Credit question 1 – Answer

The enzyme will only work with one substrate. **or**
Maltase will act on maltose and nothing else.

You can answer this question in a general way as in the first answer. You can also be more precise in your answer as in the second answer, but remember that if you make a factual error you will lose the mark.

Credit question 2

During an investigation to find the optimum temperature of an enzyme, reaction tubes containing the enzyme and its substrate were kept at different temperatures. Explain what is meant by the term optimum. (1)

Credit question 2 – Answer

The temperature at which the enzyme is most active. **or**
The temperature at which the enzyme works best.

Look out for

The term **optimum** can be used when referring to a range of factors, such as temperature and pH. The rate of an enzyme controlled reaction may be affected by either one of these factors.

Investigating aerobic respiration

What you should know at **General** level...

Cells need energy to function. The cells use energy for many purposes including:

- cell growth
- cell division
- carrying out chemical reactions
- movement.

Cells obtain their energy through the process of **aerobic respiration**. Food molecules, usually glucose sugar, are broken down in the cell to release the energy they contain. Oxygen is also needed for this to happen.

Aerobic respiration can be represented by the word equation below.

glucose + oxygen → carbon dioxide + water + energy

The carbon dioxide produced during aerobic respiration has come from the breakdown of food molecules such as glucose. Some of the energy released may be in the form of heat. This can be important in maintaining a high body temperature.

General question 1

The diagram shows the apparatus used to produce large numbers of bacterial cells for manufacturing insulin.

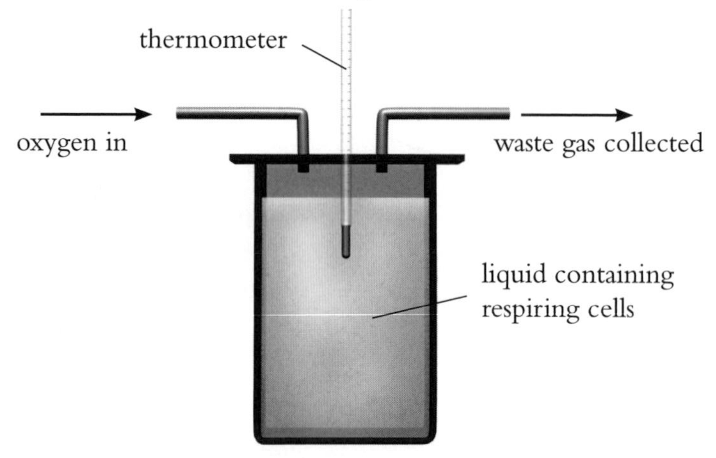

(a) Which type of respiration will take place because of the presence of oxygen? (1)

(b) What additional factor, not shown in the diagram, must be supplied to allow the bacteria to respire? (1)

(c) What waste gas will be produced during respiration? (1)

(d) What form of energy, other than chemical, may be released by the bacterial cells during respiration? (1)

General question 1 – Answer

(a) *Aerobic*

(b) *Food*

(c) *Carbon dioxide*

(d) *Heat*

Look out for

(a) Remember, all organisms will use aerobic respiration if oxygen is available. Anaerobic respiration takes place only if oxygen is not present.

(b) Respiration is the release of energy from food. Food must be present before either aerobic or anaerobic respiration can take place.

(c) All food molecules contain carbon. When the food is broken down during aerobic respiration, the carbon is converted to carbon dioxide.

(d) Some of the energy released during respiration will always be in the form of heat. This is often lost to the surroundings but it may be important in maintaining a high body temperature, as in ourselves. In manufacturing processes such as the one in this question, cooling may be needed to prevent the heat denaturing the enzymes involved.

What you should know at

level...

Fats and oils contain more energy than other food molecules. One gram of fat contains about twice as much energy as one gram of either carbohydrate or protein.

All cell reactions that build up large molecules from smaller ones (synthesis) require energy. Therefore, the manufacture of proteins and other large molecules that are produced in cells is dependent on the energy released during aerobic respiration.

Credit question 1

(a) Give one reason, other than providing heat, why cells need energy from food. (1)

(b) Which component of food provides the most energy per gram? (1)

Credit question 1 – Answer

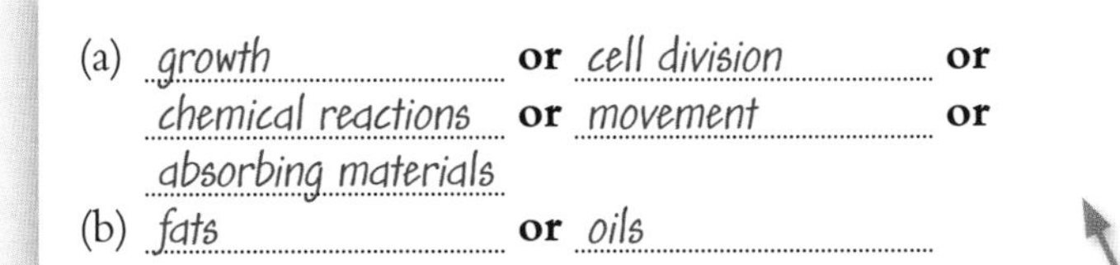

(a) *growth* **or** *cell division* **or** *chemical reactions* **or** *movement* **or** *absorbing materials*

(b) *fats* **or** *oils*

The answer must refer to energy use by a cell, not by a whole organism. Movement is acceptable because individual cells do move but an answer that refers to the movement of an organism, such as running, will lose the mark.

Chemical reactions is an acceptable answer because many reactions do require energy but naming a specific reaction which releases energy rather than uses it will lose the mark.

Respiration is a common wrong answer to this question.

We usually say that the food used in respiration is glucose, which is a carbohydrate. This is true, but other food components also contain energy and can be used for respiration. Fats and oils contain about twice as much energy as carbohydrates or proteins.

Movement

What you should know at **General** level...

The **skeleton** is made of **bones** and has the following functions:

- It provides a framework for muscle attachment.
- It provides support.
- It protects organs such as the heart, lungs, brain and spinal cord.

Bones are composed of flexible fibres and hard minerals. These two components can be seen by:

- roasting one sample of bone to burn away the fibres and leave the hard minerals.
- soaking another sample in acid to dissolve the minerals and leave the flexible fibres.

Where bones meet, a joint is formed:

- a **hinge joint** (found, for example, in knee and elbow) allows movement in one plane only.
- a **ball and socket joint** (found, for example, in shoulder and hip) allows all round movement in many planes.

The table gives information on parts of a joint and their functions.

Part of Joint	*Function*
Ligament	Holds the bones of a joint together
Cartilage	Acts as a shock absorber and reduces friction
Tendon	Attaches muscle to bone

Muscles work by contracting and pulling on bones. This causes the bone to move.

General question 1

Complete the table by giving the name of one organ that is protected by each named part of the skeleton. (2)

General question 1 – Answer

Part of Skeleton	*Organ being protected*
Backbone	*Spinal cord*
Skull	*Brain* **or** *Eyes* **or** *Ears*
Rib cage	*Heart* **or** *Lungs*

Only one answer is required for each line of the table, but the alternatives are given here as they are all acceptable answers. Remember – there is no point in giving more than one answer on each line. If one happened to be wrong, you would lose the mark. As this is a two mark question for three answers, one mark will be awarded if either only one or two of the answers is correct.

General question 2

Two separate bones were treated as described below and the results noted.
Bone 1: The bone was roasted for some time and was found to be hard and brittle.
Bone 2: The bone was soaked in acid overnight and became soft and flexible.
Complete the table to show which components of the bones had been removed by each of the treatments. (1)

General question 2 – Answer

Bone	*Component removed by treatment*
1	Fibres **or** Protein
2	Minerals **or** Calcium phosphate

In order to answer this question, you must know the substances which make up bone and apply your knowledge to the situation given. When asked for a 'component' it means a part of the whole item.

General question 3

Decide if each of the following statements about joints is **TRUE** or **FALSE** and tick the appropriate box.
If you tick the **FALSE** box you must write the correct word or words in the correction box to replace the word(s) underlined in the statement. (3)

General question 3 – Answer

Statement	*True*	*False*	*Correction*
Cartilage reduces friction at a joint.	✓		
A hinge joint allows movement in every plane.		✓	one plane only
Ligaments attach muscles to bone.		✓	Tendons

This is a popular style of question and is awarded one mark per line. Read and follow the instructions carefully when completing the table.
In this case the first statement is TRUE and only requires a tick in the TRUE box. The other 2 statements are FALSE. If you tick the FALSE box you must put a correction of the underlined words in the final column to gain the mark. ***Only*** *replace the words that are underlined.*

What you should know at Credit level...

Bones are formed by living cells. The diagram shows the structure of a synovial joint.

The tendons which attach muscles to bones are inelastic. This is so that all the force of the contraction of the muscle is transferred to the bone to bring about movement.

Each joint is operated by a pair of muscles. As one muscle contracts to move the bone, the other muscle relaxes. To move the bone back again, the opposite happens.

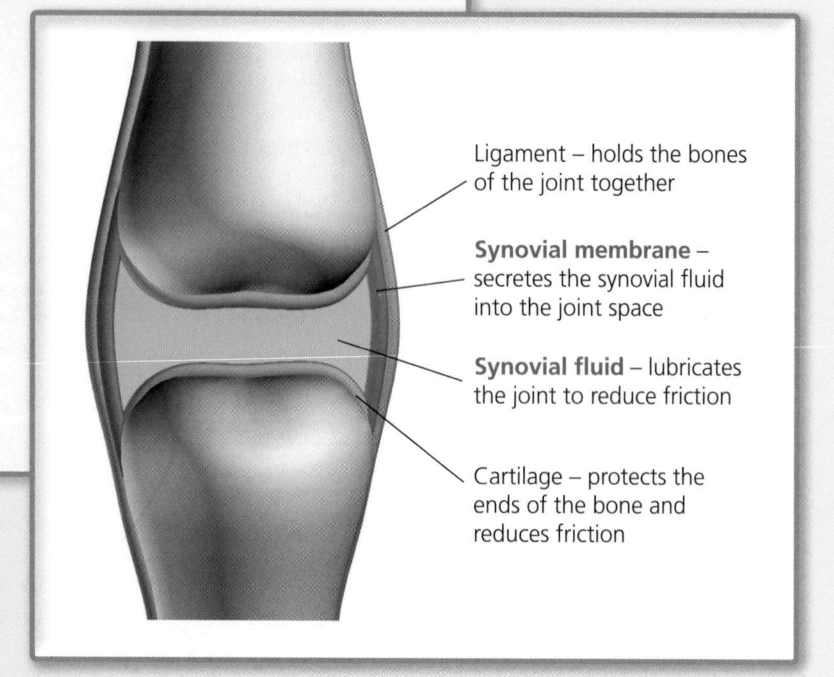

Credit question 1

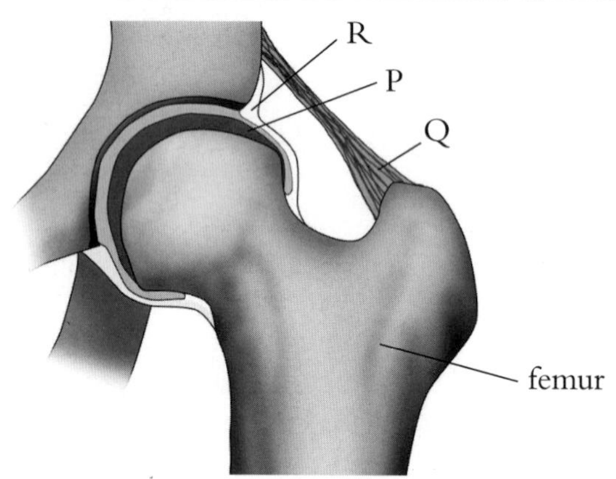

Complete the table below by inserting the correct letters, name and functions. (2)

Credit question 1 – Answer

Letter	Name	Function
R	synovial membrane	produces synovial fluid
Q	ligament	joins bones together
P	cartilage	acts as a shock absorber

When filling in a table, take care to look at the column headings and place the answer in the correct column. Don't assume that labels will be in the same positions as the diagrams you are familiar with from your school notes or books.
You need all 5 boxes correct to get the full 2 marks. 3 or 4 correct boxes will get 1 mark.

Credit question 2

Name the part of the joint which produces synovial fluid and describe the function of the fluid. (1)

Credit question 2 – Answer

Produced by *synovial membrane.*
Function *lubricates joint* **or** *reduces friction*

Be careful not to say 'prevents friction'. There is a difference between reducing and preventing something. Friction in the joint ***cannot*** *be prevented, but can be made less.*
You need ***both*** *parts of this answer to gain the mark.*

The need for energy

What you should know at General level...

Energy input and energy output must be equal if the body weight is to remain constant.

If the energy input is greater than the output, the body will gain weight.

However, if the energy input is less than the output, the body will lose weight.

The purpose of **breathing** is to carry out gas exchange between the blood and the air.

- Oxygen is absorbed into the blood from the air.
- Carbon dioxide is released from the blood into the air.

Air breathed in travels down the windpipe (**trachea**) and into the two **bronchi** which lead into the two **lungs**. In the lung, each bronchus divides into smaller tubes called **bronchioles** which continue to divide into smaller and smaller tubes which eventually end in small air sacs (**alveoli**). It is here that gas exchange takes place between the air and the blood. The blood then transports the absorbed oxygen to the body tissues.

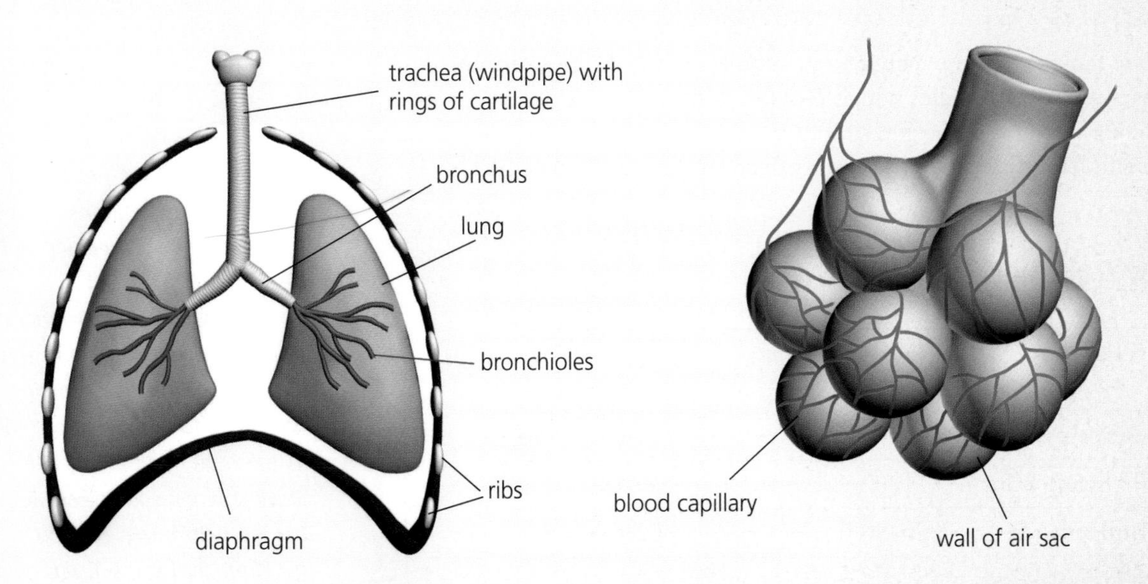

continued

What you should know at General level – continued

The function of the **heart** is to keep the blood moving round the body by continually beating. This beating can be felt as a **pulse** where the blood flows through an artery that is close to the surface.

The heart has four chambers:

- two **atria** which receive blood
- two **ventricles** which pump blood.

Valves in the heart prevent the blood from flowing backwards.

The arrows on the diagram represent the direction in which the blood flows through the heart.

Notice that the wall of the left ventricle is much thicker than the wall of the right ventricle. This is because the left side pushes blood all around the body, but the right side only has to push the blood to the lungs.

Blood flows away from the heart in **arteries** and towards the heart in **veins**. The heart itself needs a supply of blood and this is obtained through the **coronary arteries**.

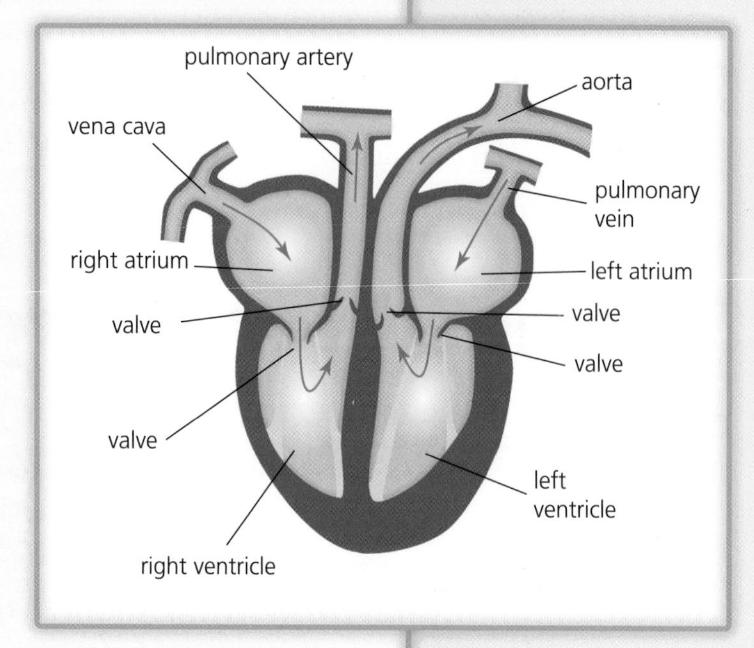

When blood rich in oxygen reaches the body cells, oxygen diffuses from the blood into the cells and carbon dioxide diffuses in the opposite direction. The blood is then returned to the lungs.

The **red blood cells** transport oxygen. The liquid part of the blood, called **plasma**, transports carbon dioxide and dissolved food, and also distributes heat around the body.

General question 1

The table shows the daily energy requirements of three people. Each person has a diet which provides him or her with 12500kJ/day for a period of four weeks.
Complete the table to show the effect on their body weight. (1)

General question 1 – Answer

Person	*Energy requirement (kJ/day)*	*Effect on body weight*
Office worker	12500	Stay the same
Pregnant woman	11000	Increase
Builder	20450	Decrease

This question requires you to compare the energy requirement in the table for each individual to the energy they gained from their food. If the intake is higher than required, the person will gain weight. If the intake is less than required, the person will lose weight.
Some questions require more than one answer to gain 1 mark. For example, in this question there are three answers to fill in the table for the mark. Never leave a blank space. As there are no half marks in Biology, it means that you will not gain anything for that question, unless it is all correct.

General question 2

Show the blood flow to and from each of the named vessels by choosing the appropriate letter from the grid. Letters may be used **once, more than once or not at all**. (3)

A Right atrium	B Left atrium
C The body	D The lungs
E Left ventricle	F Right ventricle

General question 2 – Answer

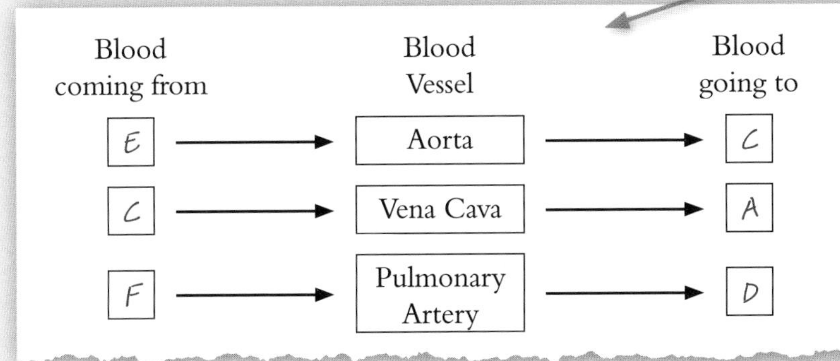

This is quite a tricky question. Try to work out the blood flow through the heart first (you can draw it out on the blank pages at the back of the exam paper), then relate the various parts to the letters in the grid.
*Blood flows **a**way from the heart in **a**rteries. Remember this by the two letter **a**'s.*
Remember to put an answer in every box, and that you can use a letter more than once or not at all as is appropriate. However, it is unlikely that one letter would be the answer to more than two parts.

General question 3

Underline the correct word in the brackets to make the sentence correct. (1)

General question 3 – Answer

The beating of the heart can be felt as a pulse. This is caused by blood flowing through {veins / capillaries / <u>arteries</u>}

In this type of question it is important that you only underline ONE word. If you make a mistake and need to change your underlined answer, make sure that it is obvious which answer you are indicating. More than one underlined, or if it is unclear which one you mean, will result in the mark being lost.

What you should know at Credit level...

Breathing is brought about by the movements of two sets of muscles.

To breathe in:

- **intercostal muscles** between the ribs contract, raising the ribcage.
- **diaphragm** muscle contracts, lowering the diaphragm.

These movements increase the volume of the chest cavity and lower the internal pressure. This causes air to rush in.

To breathe out:

- intercostal muscles relax, lowering the ribcage
- diaphragm muscle relaxes, raising the diaphragm.

These movements decrease the volume of the chest cavity and increase the internal pressure. This causes air to rush out.

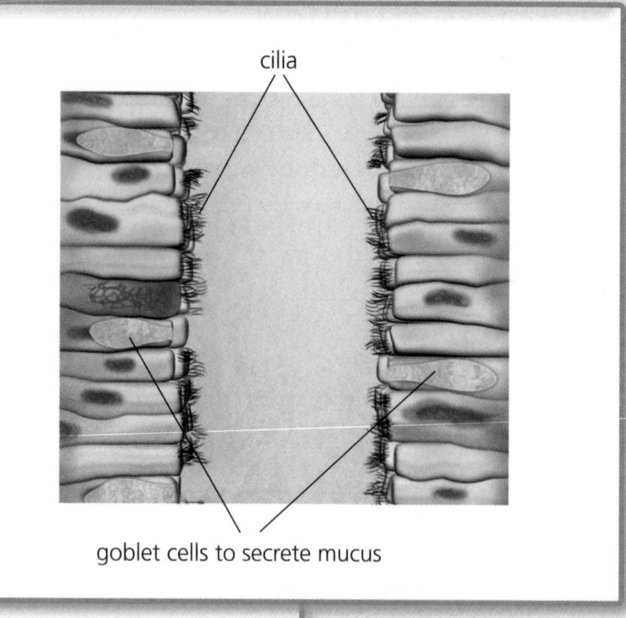

The trachea and bronchi are supported by rings of **cartilage** which prevent them from collapsing. They are also lined with cells which produce **mucus** and cells which have tiny hairs, called **cilia**, attached to them. The mucus is sticky and traps dust particles and bacteria.

The cilia beat constantly and move the mucus upwards towards the throat where it is swallowed and passes through the digestive system.

The lungs have several features which make them efficient gas exchange structures:

- a large surface area
- thin moist linings
- a plentiful blood supply.

The air breathed into the lungs has a higher concentration of oxygen than is present in the capillaries around the alveoli, so oxygen diffuses from the alveoli to the blood. As the concentration of carbon dioxide is greater in the blood than in the alveoli, carbon dioxide will diffuse in the opposite direction to the oxygen.

The red blood cells contain the chemical **haemoglobin**. This combines with the oxygen diffusing into the blood forming **oxyhaemoglobin**. The blood can carry much more oxygen as oxyhaemoglobin than it could if the oxygen was simply dissolved in the plasma.

$$\text{oxygen} + \text{haemoglobin} \rightarrow \text{oxyhaemoglobin}$$

Capillary networks in the body tissues are also efficient gas exchange structures due to the fact they too have a large surface area and thin walls. At the body tissues, oxyhaemoglobin releases the oxygen which diffuses into the cells of the tissues.

$$\text{oxyhaemoglobin} \rightarrow \text{haemoglobin} + \text{oxygen}$$

Credit question 1

Explain the function of mucus and cilia in the trachea. (2)

Credit question 1 – Answer

Mucus: *This is sticky so that dust particles and bacteria are trapped by it.*

Cilia: *These are tiny hairs which beat and move the mucus up towards the mouth.*

Be careful when describing what is trapped by mucus. 'Germs' is not an acceptable term in Standard Grade Biology although it is often used in conversation.
When describing the action of cilia, include the direction that the mucus is moved.

Credit question 2

Capillary networks allow the exchange of materials between the tissues and the blood. An example of this is oxygen moving from the lungs into the bloodstream to be taken round the body.

(a) Which component of the blood contains haemoglobin? (1)
(b) Describe the role that haemoglobin plays in the transport of oxygen. (1)
(c) Describe a feature of a capillary network which makes it efficient at exchanging gases, such as oxygen. (1)

Credit question 2 – Answer

(a) *Red Blood Cells*
(b) *Haemoglobin joins with oxygen to form oxyhaemoglobin*
(c) *It has a large surface area* **or** *the walls are very thin* **or** *it has moist surfaces*

*(a) You must state that it is the **Red** Blood Cells – Blood Cells is not enough.*
*(b) Make sure that you mention the term **oxyhaemoglobin** to show that there has been a change in the chemical once oxygen has attached to it.*
*(c) You only require one of the answers given. Make sure that you are clear about the description you give – it is not enough to say '**they** are very thin' – you must clearly state that it is the **walls** which are thin.*

Co-ordination

What you should know at General level...

The diagram shows the structure of an **eye**.

The table shows the functions of the important parts of the eye.

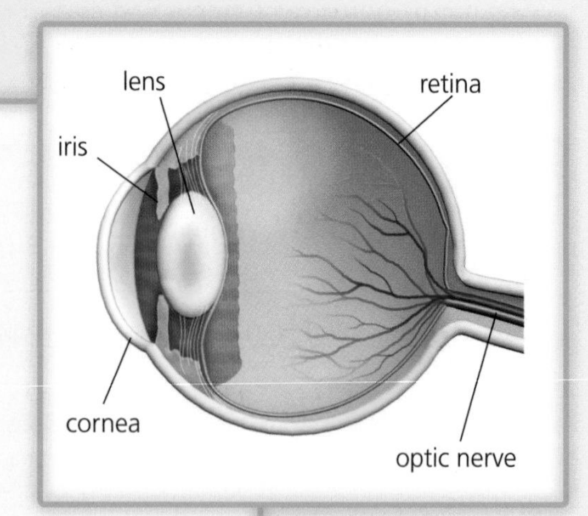

Part of eye	*Function*
Cornea	Allows light to enter and begins to focus it
Iris	Controls the amount of light entering the eye
Lens	Focuses light onto the retina
Retina	Converts light into nerve impulses
Optic nerve	Carries nerve impulses from retina to brain

Judgement of distance is more accurate using two eyes rather than one.

The diagram shows the structure of an **ear**.

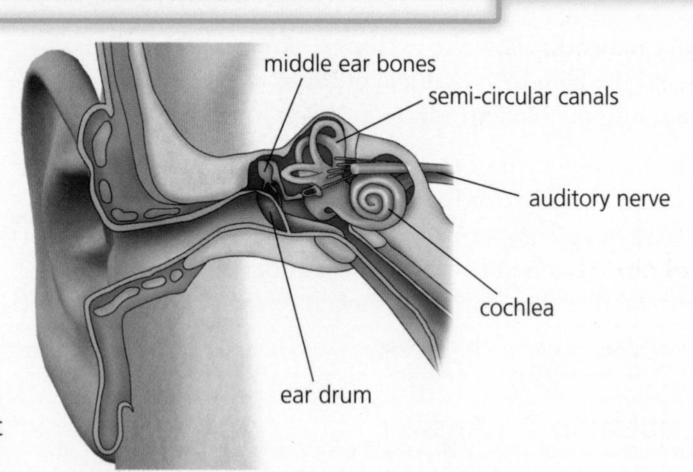

The table shows the functions of the important parts of the ear.

Part of ear	*Function*
Ear drum	Passes sound vibrations to middle ear bones
Middle ear bones	Transmits and amplifies sound vibrations
Cochlea	Converts sound vibrations into nerve impulses
Semi-circular canals	Detects movements of the head
Auditory nerve	Carries nerve impulses from cochlea and semi-circular canals to brain

Judgement of direction of sound is more accurate using two ears rather than one.

The **nervous system** is composed of the:

- **brain**
- **spinal cord**
- **nerves**.

The brain and spinal cord together make up the **central nervous system** (CNS). Nerves carry information from the senses to the central nervous system and from the central nervous system to the muscles.

General question 1

Which part of the ear is responsible for:
(a) passing vibrations to middle ear bones
(b) changing vibrations into nerve impulses? (1)

For these types of question you must be able to recall the various parts of each system. There is no alternative but to revise them until they are memorised.

General question 1 – Answer

(a) Eardrum (b) Cochlea

General question 2

The diagram shows some of the structures of the human eye.
Complete the table to give the letters, names and functions of the labelled structures. (3)

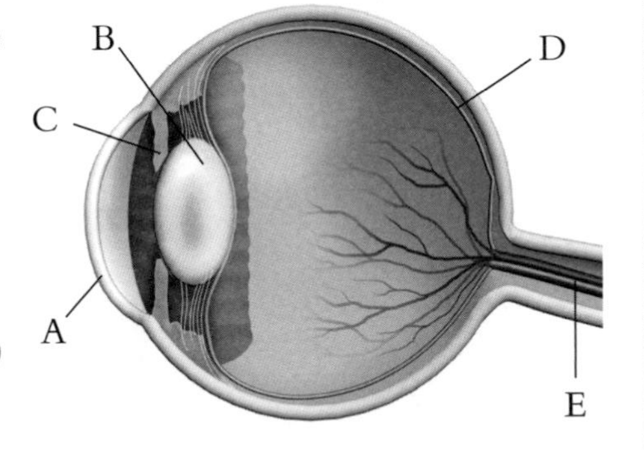

General question 2 – Answer

Letter	*Name*	*Function*
A	Cornea	Allows light to enter the eye
B	Lens	Focuses the light
C	Iris	Controls light entering the eye
D	Retina	Converts light into electrical impulses
E	Optic Nerve	Carries nerve impulses to the brain

General question 3

Choose from the list below the three parts which make up the nervous system. (1)
List
head heart nerves muscle skin lungs brain spinal cord

General question 2 – Answer

nerves spinal cord brain

General question 4

Explain the advantage to humans of having
(a) two eyes and
(b) two ears, rather than only one of each. (2)

General question 4 – Answer

(a) *It is easier to judge distances.*
(b) *It is easier to judge the direction of sound.*

In question 4 it is important to state that it is the ***distance*** *of an object and* ***direction*** *of sound that is* ***more easily judged*** *and not to say that you can see or hear better.*

What you should know at Credit level...

It is easier to judge distances with **binocular vision** (using two eyes) because each eye sees a slightly different image. The two different images which form on the retina are sent as nerve impulses to the brain, where they are 'joined together' to produce a three dimensional picture.

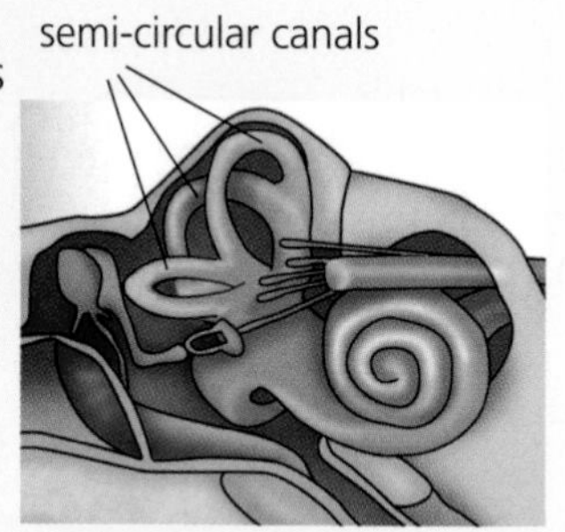

The three semi-circular canals are arranged in different planes at right angles to each other. This means that all movements of the head can be detected and the information sent as nerve impulses to the brain. This allows the control of muscles of the body so that balance and co-ordinated movement are possible.

The central nervous system (consisting of the brain and spinal cord) sorts out information from the senses and sends nerve impulses to the muscles which make the appropriate responses.

Sometimes, as a safety mechanism for the body, the **response** to a **stimulus** needs to be so fast that it does not require conscious thought by the brain. This type of response is known as a **reflex action**. A reflex action involves a series of nerve cells which form a **reflex arc**. The diagram represents a reflex arc showing:

- a **sensory nerve cell** which carries information from the senses to the spinal cord.
- a **relay nerve cell** which carries information through the CNS, passing it from sensory to motor nerve cell.
- a **motor nerve cell** which carries information to the muscles.

The brain is informed of these events but it is not involved in producing the response.

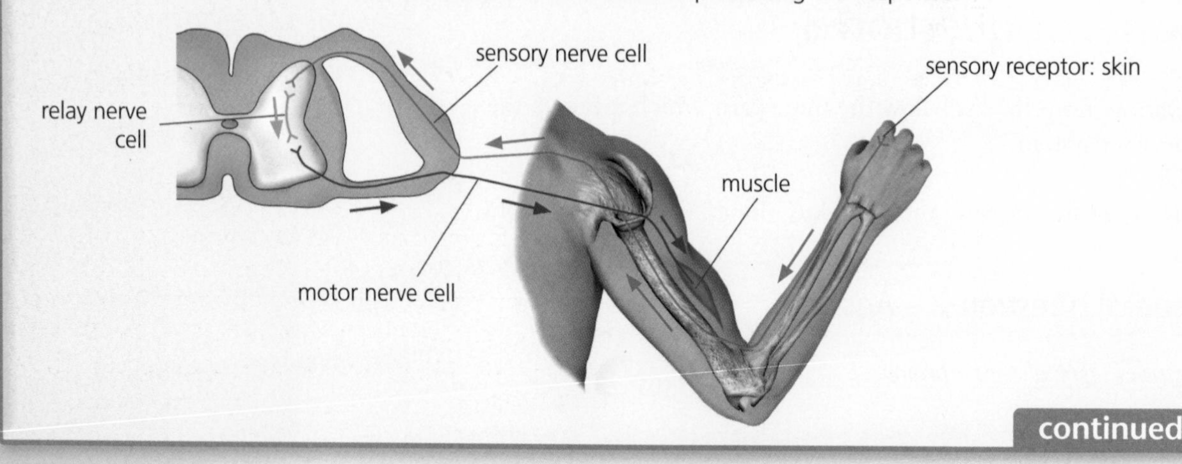

continued

What you should know at Credit level – continued

The diagram and table show the parts of the brain and their functions.

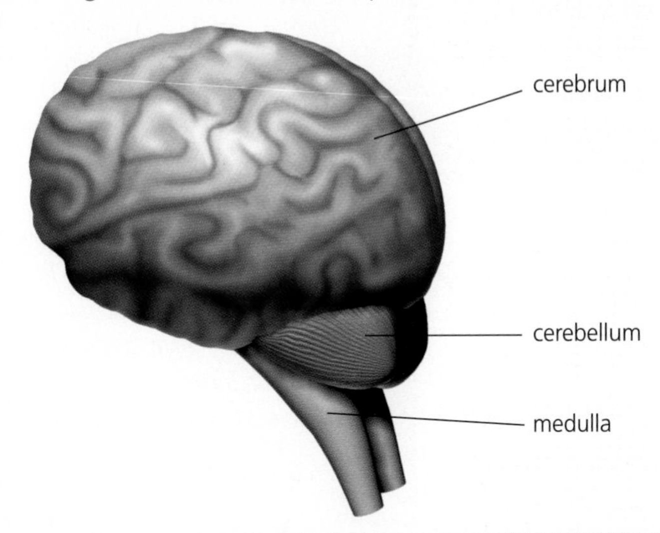

Part of brain	*Function*
Cerebrum	Controls conscious thought, memory and reasoning
Cerebellum	Controls co-ordination & balance
Medulla	Controls breathing rate & heartbeat

Credit question 1

The semi-circular canals detect movements of the head. Describe how they are arranged and explain how this helps in their function. (2)

Credit question 1 – Answer

Description: *They are arranged in three different planes.*
Explanation: *They can detect movement of the head in any direction.*
or *All movements of the head can be detected.*

Look out for

It is important to know that movement of the head in any direction or all movements of the head can be detected. If the semi-circular canals were not arranged at right angles to each other, covering three different planes, then some head movements might not be detected and balance might not be kept.

Credit question 2

A reflex action can occur to protect the body from damage. The stages involved are listed below, but not in the correct order. Draw lines between the two lists to show the order in which the events take place. (2)

Credit question 2 – Answer

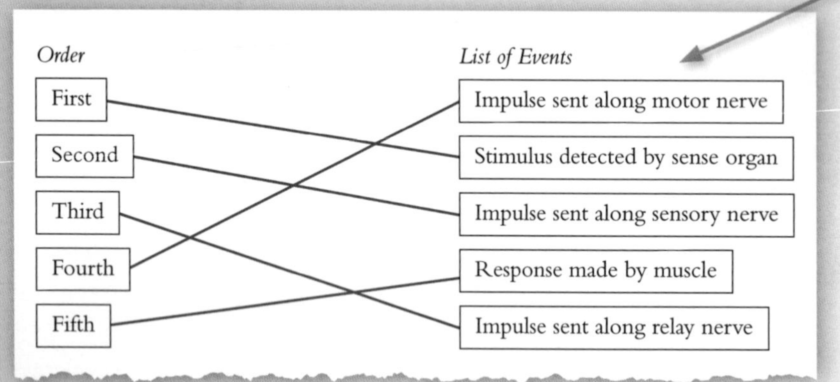

With this type of question, it is important that you are very clear about where the lines are going. Use a ruler to draw them and if you change your mind about any of them, make sure that your intention is obvious as to which is your actual answer.

Changing levels of performance

What you should know at **General** level...

Continuous or rapidly repeated contraction of muscle results in **fatigue**.

This is due to:

- a lack of oxygen, leading to **anaerobic respiration** in the muscle cells.
- a build up of **lactic acid**.

During exercise, **breathing rate** increases. This happens in order to increase the amount of oxygen getting to the muscles for **aerobic respiration**.

Pulse rate also increases as the heart pumps the blood round the body faster. This helps to increase the supply of oxygen and also glucose to the muscles. It also speeds up the removal of wastes such as carbon dioxide from them.

When exercising, breathing rate, pulse rate and lactic acid level all rise less in an athlete than in an untrained person.

The time taken for breathing rate, pulse rate and lactic acid level to return to normal after exercise is known as the **recovery time**. A person who is physically fit will have a shorter recovery time than an unfit person. This can be worked out by measuring, for example, the pulse rate before exercise and regularly after the exercise to find out how long it takes to return to normal.

General question 1

Several athletes carried out the same type of exercise for the same length of time. At the end of the exercise period their recovery times were measured. Describe how their recovery times could be used to discover who was the fittest in the group. (1)

General question 1 – Answer

The fittest person will have a faster recovery rate. **or**
The fittest person will have a shorter recovery time.

Take care over the use of the terms such as fastest, shortest, quickest and so on when describing recovery time. Make sure that the phrase you choose means that the fittest person has their pulse rate and breathing rate return to their own normal rate first.

General question 2

The table contains factors which may change during strenuous exercise. Complete the table by writing *increases*, *decreases*, *stays the same*, in the appropriate boxes.
Each phrase may be used **once**, **more than once**, or **not at all**. (2)

General question 2 – Answer

Factor	*During Exercise*
Breathing rate	*increases*
Pulse rate	*increases*
Carbohydrate store	*decreases*
Lactic acid production	*increases*

General question 3

Complete the following sentence about the causes of muscle fatigue. (2)

General question 3 – Answer

Muscle fatigue is caused by a build up of *lactic acid* in the muscles due to the lack of *oxygen*.

This is straight recall of knowledge and time needs to be spent learning these facts.

What you should know at Credit level...

Muscle fatigue occurs when there is not enough oxygen present for aerobic respiration to produce all the energy required. **Anaerobic respiration** then begins in order to increase energy release.

Glucose → Lactic acid + energy

The build up of lactic acid causes soreness in the muscles and they can no longer work as efficiently. Normal functioning of the muscles is restored when the lactic acid is removed in the presence of oxygen.

Training improves the efficiency of the lungs and circulation. As a person trains and becomes fitter, pulse rate does not go up so high during exercise and returns to normal faster. This means that the recovery time of a person training will become less.

Credit question 1

Name the process in the body cells which produces lactic acid and causes muscle fatigue. (1)

Credit question 1 – Answer

Anaerobic respiration.

Look out for

You need to know the differences between aerobic and anaerobic respiration in terms of the raw materials and the waste products. Learn the equations for each.

Credit question 2

(a) When an athlete trains, the efficiency of the heart muscle increases. Explain how an increase in the number of blood vessels in the heart can help to make it more efficient. (1)

(b) Name another organ of the body which shows increased efficiency with training. (1)

Credit question 2 – Answer

(a) *There is an increase in the supply of oxygen* **or** *food* **or** *There is an increase in the removal of carbon dioxide.*

(b) *Lungs*

Look out for

When an athlete trains to become fitter, the efficiency of the lungs and circulation both increase. Therefore training improves the flow of blood around the body. The delivery of essential substances and removal of waste materials is more efficient.

Variation

What you should know at **General** level...

A **species** is a group of interbreeding organisms whose offspring are fertile.

Within a species, **variations** (differences) occur and can be of two types:

- **continuous variation** – for example; height, mass, length.
- **discontinuous variation** – for example; eye colour, blood group, tongue-rolling ability.

General question 1

Complete the table by choosing words from the list to match the descriptions. (2)

List habitat species variation phenotype mutation

General question 1 – Answer

Description	*Word*
Differences which occur between organisms.	*Variation*
A group of organisms which produce fertile young when bred together.	*Species*

Always read the whole list of words given carefully, and then match the correct one to the description. Do not just pick the first one you think matches without checking all of them out first.
*Remember, the word **list** is not a possible choice.*

General question 2

In a scientific investigation, a tiger female was successfully bred with a lion. The resulting offspring was known as a liger. Scientists found that he was infertile.

(a) What piece of information given above, might suggest that the tiger and the lioness were the same species? (1)

(b) What evidence is there in the investigation, that the two animals were not the same species? (1)

General question 2 – Answer

(a) *The animals successfully bred to produce an offspring.*

(b) *The offspring was infertile.*

In this question the knowledge you have about species must be applied to the specific situation as it is described. Care needs to be taken to pick out only the relevant parts of the information for the answer.

What you should know at Credit level...

Continuous variation shows a range of values between a minimum and a maximum value. Examples include height of trees, handspan, diameter of limpets shells. This type of variation is often represented by a line graph.

Discontinuous variation can be put into distinct categories or groups. Examples include eye colour, shoe size, leaf shape. This type of variation is often represented by a bar graph.

Credit question 1

Variation exists between members of the same species.
Explain the term 'continuous variation'. (1)

Credit question 1 – Answer

A characteristic which shows a range of values between a minimum and a maximum value.

Credit question 2

Using eye colour as an example, explain what is meant by 'discontinuous variation'. (1)

Credit question 2 – Answer

There are a limited number of distinct eye colours.

These questions are usually straightforward and if you have learned the definitions of continuous and discontinuous variation, you should have no trouble in picking up the marks for them. The key point is whether or not there is a continuous range of possibilities.

Credit question 3

Is variation in blood groups an example of continuous or discontinuous variation?
Explain your answer. (1)

Credit question 3 – Answer

Discontinuous.
Explanation: *There are a limited number of possibilities and individuals can be put into distinct groups.*

What is inheritance?

What you should know at General level...

Certain characteristics are determined by genetic information received from the parents.

Examples include eye colour, tongue-rolling ability, presence of ear lobes, shape of flowers, size of leaves, colour of tree bark.

Each body cell has two matching sets of **chromosomes** which code for the characteristics. One set has come from the male parent and the other set from the female parent. The chromosomes are carried in the sex cells of the parents and so they combine at fertilisation.

A part of a chromosome which codes for a characteristic is called a **gene**. A characteristic is controlled by two forms of a gene.

The set of genes possessed by an organism is known as its **genotype**. (Remember this as they both begin with the same three letters.)

The appearance of the organism which results from its genotype is known as the **phenotype**. For example, for the gene controlling eye colour, a person may have the phenotype of brown or blue or green etc.

In a genetic cross, abbreviations are used to identify different generations.

P = Parents
F_1 = First filial generation (1st generation of offspring)
F_2 = Second filial generation (2nd generation of offspring)

The following example of a genetic cross is between a tall plant and a dwarf plant. The tall gene (**T**) is **dominant** over the dwarf gene (**t**) which is **recessive**. (The dominant gene shows its effect in the phenotype of an organism and masks the presence of a recessive gene.)

P	(phenotype)	Tall	×	Dwarf
	(genotype)	TT	×	tt
	(gamete genotypes)	T	×	t
F_1	(genotype)		Tt	
	(phenotype)		All Tall	
F_2	($F_1 \times F_1$ genotypes)	Tt	×	Tt
	(gamete genotypes)	T and t	×	T and t

1st parent gamete genotypes (columns); 2nd parent gamete genotype (rows); inner cells = F_2 possible genotypes

	T	t
T	TT	Tt
t	Tt	tt

F_2 genotype ratio 1TT : 2Tt : 1tt

F_2 phenotype ratio 3 Tall : 1 Dwarf

continued

What you should know at General level – continued

Genotypes which have two identical forms of a gene are known as **true-breeding**. Any individual in the above cross with the genotypes **TT** or **tt** are true-breeding. Examples are both parents, and the F_2 individuals which have the genotypes **TT** and **tt**.

The individuals in the $\mathbf{F_1}$ are uniform in that they all show the same phenotype and it is the dominant one. These individuals are not true-breeding because they also carry the recessive gene.

Sex cells are called **gametes**. When gametes are formed there is a reduction to a single set of chromosomes. Each sex cell carries one complete set of chromosomes, so each parent contributes one of the two inherited forms of a gene at fertilisation.

The sex of a child is determined by specific chromosomes called **X** and **Y chromosomes**. In humans, the male gamete (sperm) may have an **X** or a **Y** chromosome, while each female gamete (egg) has an **X** chromosome.

At fertilisation, a sperm nucleus joins with an egg nucleus, bringing together either an **X** chromosome from the mother and a **Y** chromosome from father (**XY** = male baby) or an **X** chromosome from the mother and an **X** chromosome from father (**XX** = female baby).

General question 1

Seed colour in pea plants is controlled by genes.

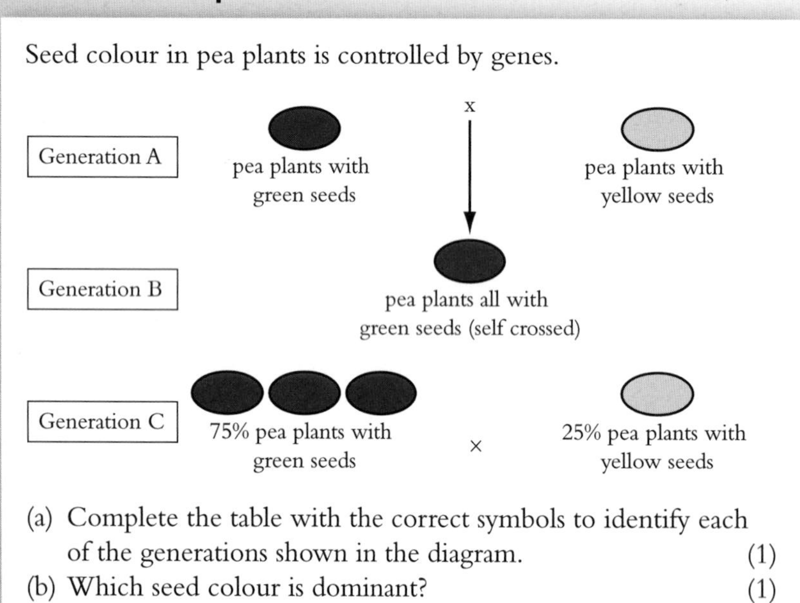

(a) Complete the table with the correct symbols to identify each of the generations shown in the diagram. (1)

(b) Which seed colour is dominant? (1)

General question 1 – Answer

(a)

Generation	Symbol
A	P
B	F_1
C	F_2

(b) Green

(b) To tell which is dominant, look for the characteristic which shows in the F_1 generation (generation B in this case) and of which there are more in the F_2 (generation C).

General question 2

The following table gives a list of family members.
Opposite each, write **XX** or **XY** to indicate the sex chromosomes carried by each person. (1)

General question 2 – Answer

Member of Family	*Sex Chromosomes*
Son	XY
Mother	XX
Nephew	XY
Father	XY

*You need to remember that any male person will carry an **X** and a **Y** chromosome, while all females will carry two **X** chromosomes. (Hint: think of the idea that the 'the guys wear the Y's!'.)*

General question 3

How many sets of chromosomes are found in:
(a) a sperm cell
(b) a body cell? (1)

General question 3 – Answer

(a) *1 set*
(b) *2 sets*

Look out for

Remember – different species have different numbers of chromosomes. You are not expected to know how many any species has, but you are expected to know that all species have 1 set of chromosomes in their sex cells and 2 sets in their body cells.

What you should know at Credit level...

Different forms of a gene are called **alleles**.

A **monohybrid cross** is a genetic cross in which only one characteristic is studied. The parents in experimental monohybrid crosses are usually true-breeding and show different phenotypes of the same characteristic.

The F1 generation from such crosses all show the same phenotype (the dominant one) and their genotype shows one allele from each parent.

The F2 generation shows both phenotypes but in the ratio of three of the dominant variety to one of the recessive variety if the original parents were true-breeding.

The predicted ratios of any cross are rarely achieved accurately. This is due to the fact that fertilisation is a random process involving the element of chance. It may also be due to the fact that in some cases the sample size was too small.

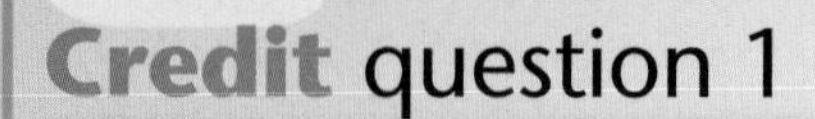

Credit question 1

The following cross was carried out using two varieties of the fruit fly, *Drosophila*. One type had long wings and the other had stunted wings.

P Long wings × Stunted wings

$\mathbf{F_1}$ All long wings

F_1 × F_1

$\mathbf{F_2}$ Some long wings and some stunted wings

(a) Complete the table using the generation symbols **P**, $\mathbf{F_1}$ and $\mathbf{F_2}$ in the appropriate space to indicate whether true breeding individuals would be present. (1)

(b) Predict the expected ratio of long winged flies to stunted winged flies in the F_2 generation. (1)

(c) The predicted ratio was not achieved in the actual cross. Give a reason to account for this. (1)

Credit question 1 – Answer

(a)

Contains true breeding individuals	*Does not contain true breeding individuals*
P F_2	F_1

(b) *3 long wings : 1 stunted wings*

(c) *Fertilisation is a random process* **or** *Sample size may have been too small.*

a) The F_1 generation all have long wings and will have a long wing allele. They all must also contain the recessive allele for stunted wings from the parent with stunted wings. This means that the F_1 flies will all have a dominant and a recessive allele and will not be true-breeding.

b) You must learn the expected outcomes and ratios associated with the F_1 and F_2 generations. If the parents are true breeding to start with, the F_1 will all show the dominant form and will not be true breeding. The F_2 will show some of each form with about ¾ of the total showing the dominant feature and ¼ showing the recessive feature.

c) Either of the answers shown is acceptable in this case, but if the example in the question involves a large sample, then you must use the first answer.

Credit question 2

In Guinea pigs, black coat is dominant to cream coat. **B** represents the form of the gene for black coat and **b** represents the one for cream coat.

A cross was carried out using the parents shown below.

Parents **BB** × **bb**

(a) (i) Give the genotype of the F_1 offspring in this cross. (1)

(ii) What phenotype(s) will the F_1 show? (1)

(b) What term is used to describe different forms of the same gene such as black coat and cream coat? (1)

Credit question 2 – Answer

(a) (i) *Bb*

(ii) *All will be black.*

(b) *Allele*

Look out for

Genetics has a whole set of words with very specific meanings. It is worth taking the time to learn the definitions as they are the key to getting the marks.

Genetics and society

What you should know at General level...

Characteristics in both plants and animals can be improved through the process of **selective breeding**. This means selecting organisms which exhibit the most desirable characteristics and breeding them together.

Examples of improvement include increased yield, increased disease resistance and increased growth.

A change in the number or structure of one or more chromosomes is known as a **mutation**. In humans, a condition known as Down's Syndrome is caused by a chromosome mutation.

A technique called **amniocentesis** can be carried out before birth to detect chromosome characteristics. This involves inserting a needle into the uterus and removing a sample of the fluid surrounding the baby. The fluid contains cells from the baby which are examined for chromosome abnormalities.

General question 1

(a) Careful breeding of cattle can bring about an increase in the quantity of milk produced by dairy cows. Name another characteristic which can be improved by this method. (1)

(b) What general name is given to the procedure of crossing particular plants or animals to bring about improvement? (1)

General question 1 – Answer

(a) *Better quality wool* (sheep) **or** *larger yield* (wheat) **or** *disease resistance* (potatoes)

(b) *Selective Breeding*

(a) There are many examples here and they can be chosen from animals or plants as the question is not specific about the type of organism. It is not necessary to name the organism involved, but it is easier to come up with an answer if you have a particular organism in mind.
(b) You must be quite specific about this term – no other answer will do.

General question 2

(a) Genetic mutation can give rise to differences between individuals. Name a human condition caused by a chromosome mutation. (1)

(b) Name the technique used to sample a baby's chromosomes before birth. (1)

General question 2 – Answer

(a) *Down's syndrome*

(b) *Amniocentesis*

(a) While there are many acceptable answers to this question, it is best to learn Down's Syndrome as the answer to this type of question, as this is the most widely quoted example.
(b) As with many words in Biology, this is a difficult one to spell. If you have difficulty with the correct spelling, then first learn how to say it and then write it as you would sound it out to make the word pronounce correctly. If you do this then you will probably get the mark.

What you should know at Credit level...

The table shows examples of improvements brought about by selective breeding.

Organism	*Improvement*
Cattle	Increased milk or beef yield
Potatoes	Increased disease resistance
Sheep	Increased wool quality

Some chromosome mutations can be advantageous and produce individuals which are in some way better than the original. This can be of economical importance. Mutations can cause larger fruit to be produced (strawberries, apples) or in some cases can give resistance to disease both in plants and in humans.

Substances which increase the rate of mutation are called **mutagenic agents**. Examples are high temperatures, certain chemicals (for example, mustard gas) and radiation (for example, X-rays).

Credit question 1

Beef quality in cattle can be improved by selective breeding. Name a different organism which can be improved by selective breeding and describe the improved characteristic. (1)

Credit question 1 – Answer

Organism: *Sheep* **or** *Corn* **or** *Tomato*
Improvement: *better quality meat* **or** *larger yield of oil* **or** *disease resistance*

*There are many examples that can be chosen from animals or plants. It would, however, not be acceptable to give **cattle – improved milk yield**, as the question asks for a **different** organism to the one given. It is very important to read the entire question carefully. It would be safer to choose an example that you know has commercial importance.*

Credit question 2

(a) What is meant by the term 'mutation'? (1)
(b) Give an example of a factor which can increase the rate of mutation. (1)
(c) Give an example of a mutation which is advantageous in plants. (1)

Credit question 2 – Answer

(a) *A change in the number of chromosomes* **or** *A change in chromosome structure* **or** *A change in genetic information.*
(b) *Radiation* **or** *UV light* **or** *X-Rays* **or** *High temperatures* **or**. *Chemicals*
(c) *Hardier plants* **or** *Increased yield* **or** *Increased resistance to disease.*

(a) Remember, a mutation is not an individual organism which has an abnormality. Your answer must refer to the genetic information within the cells. There is a wide variety of answers to these questions. Try to focus on one or two examples of each answer and learn those.

Living factories

What you should know at **General** level...

Biotechnology refers to the use of micro-organisms for the benefit of people.

Yeast is used in the making of beer and wine. It also causes the dough to rise when bread is made.

Yeast is a single-celled fungus which can use sugar as food for energy. In the absence of oxygen this is called **anaerobic respiration** or **fermentation**.

Fermentation of glucose by yeast produces alcohol and carbon dioxide as well as releasing some energy.

glucose → alcohol + carbon dioxide + energy

Bacteria are micro-organisms used to make cheese and yoghurt from milk.

- Bacteria can cause milk to turn sour because of anaerobic respiration or fermentation.

General question 1

(a) What type of micro-organism is yeast? (1)

(b) Yeast is important in making bread and beer through the process of fermentation.
State why yeast is required in each case. (2)

General question 1 – Answer

(a) *Yeast is a fungus.*

(b) Bread: *Makes the dough rise*
Beer: *Produces alcohol or ethanol*

(b) It is not sufficient to say that carbon dioxide is produced without stating the effect that the carbon dioxide has on the dough or bread.
It is not sufficient to say that the yeast ferments the sugar as this is mentioned in the question.

General question 2

(a) Name the type of micro-organism used in the manufacture of yoghurt from milk. (1)

(b) Explain why containers are sterilised before being used for making yoghurt. (1)

(c) Micro-organisms carry out fermentation of the sugars in milk.
What effect does this have on the milk? (1)

General question 2 – Answer

(a) *Bacteria.*
(b) *To prevent contamination from other unwanted micro-organisms.*
(c) *The milk turns sour.*

(b) The key phrase in this question is 'before being used'. Sterilisation doesn't stop other micro-organisms entering the container but it does kill them once they are there.
(c) You must describe a change in the milk as a result of the fermentation. It is not sufficient to say that acid is produced without stating the effect that the acid has.

What you should know at Credit level...

Aerobic respiration and anaerobic respiration are two processes which release energy from food. The two processes have some similarities and some differences as shown in the table.

Respiration	*Organisms*	*Oxygen*	*Reactants*	*Products*
aerobic	all	present	glucose + oxygen	water + carbon dioxide + energy
anaerobic	plants and fungi	absent	glucose	alcohol + carbon dioxide + energy
	animals and bacteria	absent	glucose	lactic acid + energy

During the manufacture of beer, brewers must provide the correct conditions for fermentation.

- The fermentation vessel has to be sterilised to remove unwanted micro-organisms. The sterilisation process must destroy bacterial and fungal **spores** which can withstand relatively high temperatures such as boiling water so temperatures in excess of 120°C are used.
- Food for the yeast is supplied by the addition of maltose from malted barley.
- Fermentation also produces heat energy which could prevent enzymes in the yeast from working at their best, so the fermentation vessel often has to be cooled to the optimum temperature.
- Oxygen is prevented from entering the fermentation vessel.

Beer is manufactured by **batch processing**. This means that:

- All the ingredients are added to the fermentation vessel at the start of the process and are left for the required length of time.
- When the process is complete the product (beer) is removed and the fermentation vessel is cleaned and sterilised ready to be used again.

Barley grains (seeds) contain starch. The starch cannot be used by yeast as a food source. During the germination of barley seeds, starch is converted into the sugar maltose.

Malting is a process in which barley seeds are germinated thus converting starch into **maltose**. The germination is then stopped and the malted barley now contains a good supply of maltose sugar which can be fed to the yeast to make beer.

The sugar in milk is called **lactose**. Bacteria can ferment this sugar producing lactic acid which lowers the pH and turns the milk sour.

Credit question 1

Industrial processes such as beer-making can be carried out in large fermenters. After fermentation is complete, the fermenter is drained and the useful product is separated.
New starting ingredients can then be added to the fermenter.

(a) What name is given to this type of process? (1)

(b) When the vessel is empty, it is treated to destroy residual spores of fungi and bacteria. How could this be done? (1)

Credit question 1 – Answer

(a) *Batch process.*

(b) *The vessel should be heated to at least 120°C or cleaned with disinfectant.*

It is not correct to just answer 'heat it to a high temperature'. 100°C is a high temperature but it is not enough to kill bacterial spores. Your answer must also be practical. The question refers to an industrial fermenter so it would not be possible to put it into a pressure cooker.

Credit question 2

Why must the barley be malted before it can be used by the yeast? (1)

Credit question 2 – Answer

Yeast cannot ferment starch, malting converts the starch to sugar **or** *maltose.*

If you name the sugar then it must be maltose and not glucose.

Problems and profit with waste

What you should know at General level...

The disposal of untreated **sewage** into rivers and seas can be damaging to living organisms and the environment.

- Fish and aquatic invertebrates may die due to lack of oxygen.
- Humans can catch diseases such as dysentery, typhoid and cholera.

During sewage treatment, harmful components in sewage are broken down by decay micro-organisms into harmless products which can be safely released into the environment. To do this, the decay micro-organisms need a plentiful supply of oxygen which can be provided in different ways:

- compressed air can be bubbled through it
- paddles can stir the liquid
- the liquid can be trickled over stones in a gravel bed.

continued

What you should know at General level – continued

Micro-organisms can cause disease. When working in a laboratory, scientists use safety precautions to ensure that micro-organisms are not released into the environment and that unknown micro-organisms do not contaminate their experiments. The main safety precautions are:

- Wash hands and benches before and after work.
- Use sterile equipment.
- Dispose of used equipment only after it has been sterilised.
- Use special techniques to transfer micro-organisms from one container to another.

Similar precautions are carried out in many biotechnological industries. Any contamination with unwanted micro-organisms will spoil the product and could cause an infection in someone which could be life threatening.

Biotechnological processes can produce waste which could be harmful if it was released into the environment. If this waste could be treated to change it into a useful product, then more profit could be made and less damage would be caused to the environment.

- The fermentation of sewage sludge produces **methane** (biogas) which can be collected and used to provide power for the sewage treatment plant.
- Yeast can ferment waste sugar into **alcohol**. The alcohol can be mixed with petrol and used as fuel in cars.

Both methane and alcohol are produced by fermentation. They are renewable fuels and their use would help to conserve **fossil fuels**.

Under ideal conditions micro-organisms grow rapidly by asexual reproduction. The micro-organisms can be harvested and used as a protein rich food. Some species of bacteria can be used as cattle feed and certain fungi can be used to make **mycoprotein**.

General question 1

The diagram shows one type of sewage treatment works.

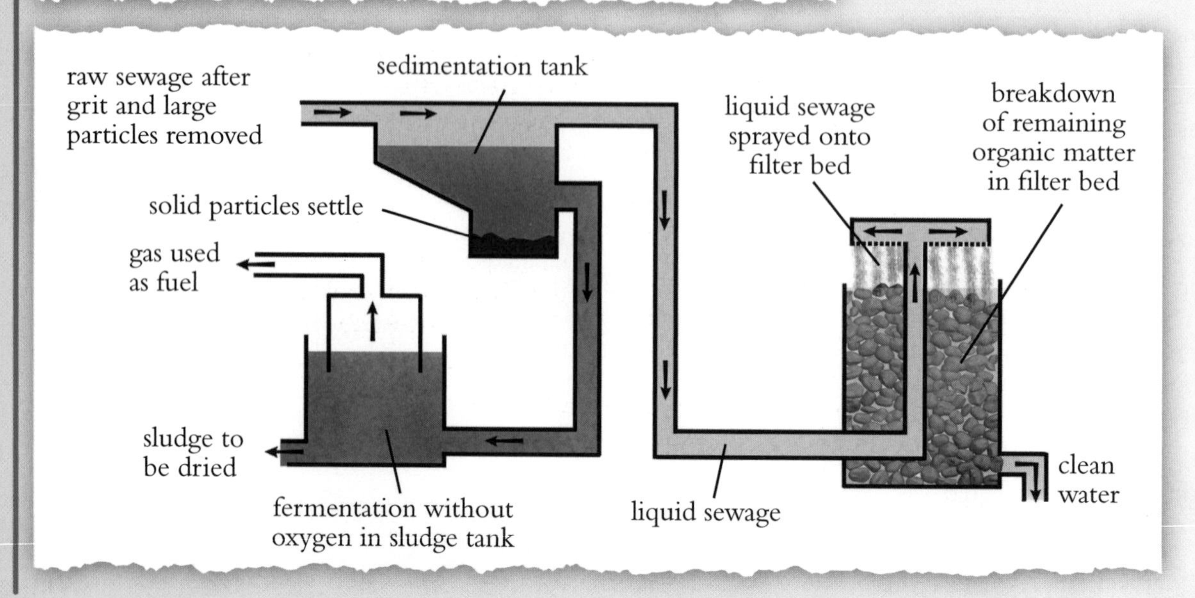

(a) What causes the breakdown of organic matter in the filter bed? (1)

(b) The filter bed contains layers of stones and gravel. How does this help to provide the oxygen needed for the breakdown of organic matter? (1)

(c) (i) Name the gas which is produced in the sludge tank and which can be used as a fuel. (1)

(ii) State one advantage of using fuels obtained by fermentation rather than fossil fuels. (1)

A wide range of micro-organisms is used. Any type would also be correct: for example, bacteria, fungi, protozoa.

General question 1 – Answer

(a) *Micro-organisms*

(b) *There are air spaces between the stones which provide oxygen.*

(c) (i) *Methane* **or** *Biogas*

(ii) *Fermentation fuels are renewable.* **or** *They reduce the use of fossil fuels.*

Look out for

Don't confuse the word renewable with re-usable – you can't re-use any fuel after it has been burnt.

What you should know at Credit level...

Many bacteria and fungi produce spores which are resistant to high temperatures and disinfectants. If they are given favourable conditions these spores can germinate causing the bacteria and fungi to grow which could cause the contamination of biotechnological processes. Temperatures in excess of 120°C are required to sterilise equipment and to kill spores.

Decay is a natural process in which organic matter is broken down by micro-organisms. The organic matter provides food for the micro-organisms and supplies them with energy. Nutrients from the organic matter such as nitrogen and carbon are recycled.

Sewage treatment depends on micro-organisms decaying the organic waste. These micro-organisms feed on the sewage and respire aerobically completely decomposing it into harmless products. If there is a lack of oxygen, the micro-organisms respire anaerobically and are only able to partially decompose the sewage.

Sewage is a complex mixture of many different organic chemicals. Individual types of micro-organisms can only decompose some of these chemicals so to ensure that all the chemicals are decomposed, a wide range of micro-organisms are needed to successfully treat sewage.

Micro-organisms can be used to **upgrade waste** from biotechnological processes changing it into products with a higher energy or protein content. Rather than being sent for disposal the upgraded the waste can now be sold for profit.

Credit question 1

(a) In a sewage works, micro-organisms cause the decay of sewage. What is the benefit to the micro-organisms in carrying out this process? (1)
(b) What type of respiration must be carried out by the micro-organisms to ensure the complete breakdown of the sewage? (1)
(c) Sewage contains a wide range of materials. What ensures that all these materials are broken down? (1)

Credit question 1 – Answer

(a) *The micro-organisms get food* **or** *energy from the sewage.*
(b) *Aerobic respiration.*
(c) *A wide range of micro-organisms are used.*

(b) You only know two types of respiration – aerobic and anaerobic.
(c) The word 'microbes' is an acceptable abbreviation for micro-organisms. The word 'bugs' is not acceptable. Examples of micro-organisms such as bacteria or fungi are also acceptable.

Credit question 2

The diagram shows part of the carbon cycle.

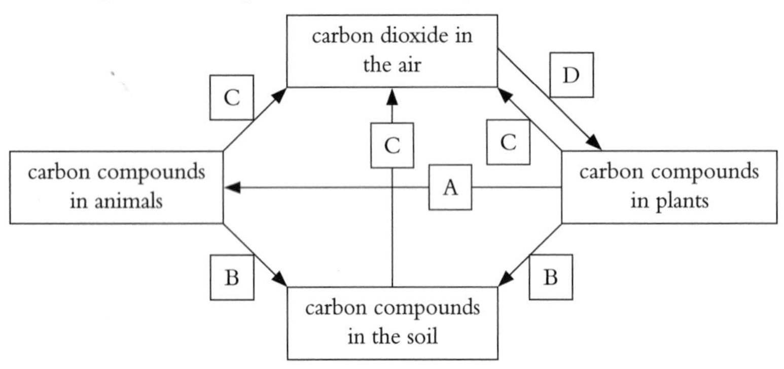

(a) Use one letter from the diagram to identify each of the stages in the table below. (2)
(b) Name a type of organism responsible for process B (1)

Credit question 2 – Answer

(a)

Stage	*Letter*
photosynthesis	*D*
death and decay	*B*
respiration	*C*

(b) *bacteria* **or** *fungi*

Look out for

Make sure you know that photosynthesis takes carbon dioxide *out* from the air and that respiration adds carbon dioxide *into* the air.

Don't try to name a specific micro-organism – you could be wrong and lose an easy mark!

Reprogramming microbes

What you should know at **General** level...

Bacteria contain chromosomal material which controls their functions. Pieces of chromosome from the cells of other organisms can be transferred into bacterial cells and allowed to combine with the bacterial chromosomal material. This new genetic material can programme the bacterial cells to make new products.

This process is known as **genetic engineering** and has been used to make useful products such as **insulin** and human growth hormone. Insulin is used to treat **diabetes** and human growth hormone is used to treat growth disorders.

Bacterial products such as enzymes can be added to detergents to make '**biological detergents**'.

The growth of bacteria can be prevented by using **antibiotics** which are used to treat infections.

General question 1

(a) Diabetes can be treated with a substance produced by genetic engineering. Name this substance. (1)

(b) What type of chemical, used in biological washing powders, can be produced by genetic engineering? (1)

(c) During genetic engineering, what is transferred into bacteria from another organism? (1)

General question 1 – Answer

(a) *Insulin*

(b) *Enzymes*

(c) *Chromosome material* **or** *genes.*

The question asks for a type of chemical. You cannot answer with the name of an enzyme. The answer 'protein' is also unacceptable. Although all enzymes are made of protein, there are many proteins which are not enzymes.

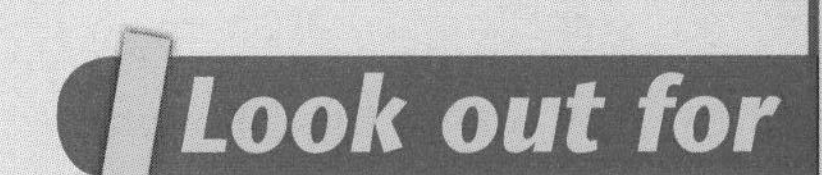

Whole chromosomes are not transferred between organisms.

General question 2 (2003 Q17)

The following diagram describes some of the stages involved in transferring a gene from a **human** chromosome into a bacterial cell.

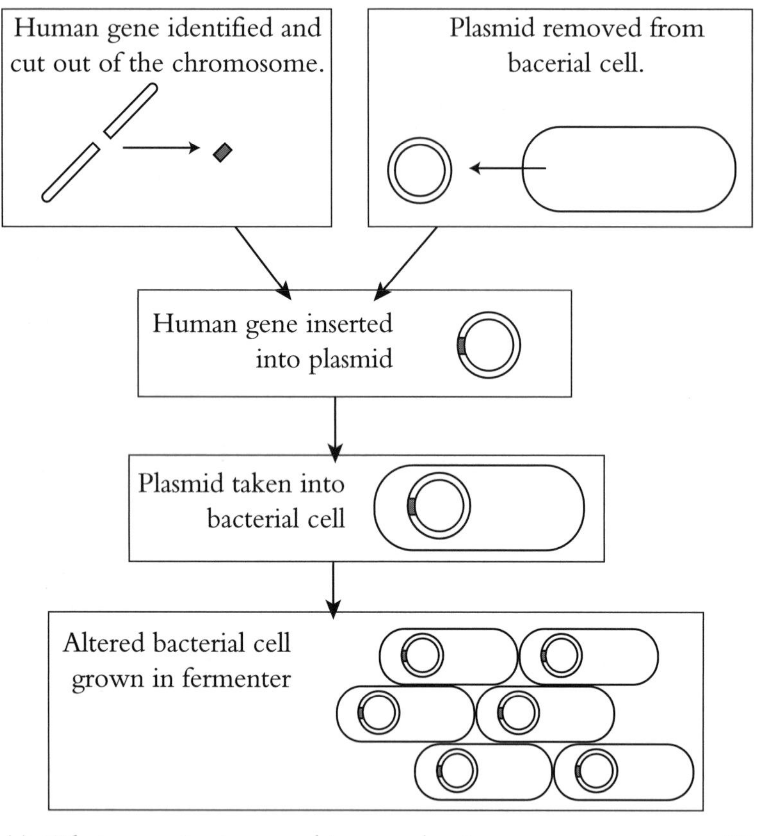

(a) What name is given to this procedure? (1)
(b) Give an example of a product that can be made by bacteria as a result of this procedure. State the use of this product. (2)
(c) What type of reproduction is involved during the growth of the bacteria in a fermenter? (1)

General question 2 – Answer

(a) *Genetic engineering* **or** *Genetic modification.*

(b) Any one of the following:
Product: *Insulin.* Use: *treat diabetes* **or** *control blood sugar levels.*
or Product: *Growth Hormone.* Use: *treat growth problems*
or Product: *Factor VIII.* Use: *treat haemophilia*
or Product: *Interferon.* Use: *treat cancer.*
Product = 1 mark Correct use = 1 mark

(c) *Asexual reproduction.*

(a) This is not cloning. Cloning is a different process.
(b) You must match the product and use. You will lose marks if you give a product and an incorrect use.
(c) You only know the names of two types of reproduction. Bacterial reproduction does not involve the joining of gametes or fertilisation so the answer cannot be 'sexual reproduction'.

What you should know at **Credit** level...

In addition to a single **circular chromosome**, bacteria usually have several small circular pieces of genetic material called **plasmids**.

Genetic engineering techniques make use of the fact that plasmids can easily be transferred from one bacterial cell to another. For example a plasmid containing the human gene for insulin can be put into a bacterial cell. The cell will copy the plasmid and pass it onto many other cells. This process will be repeated many times. The modified cells will reproduce rapidly creating many bacteria capable of making insulin.

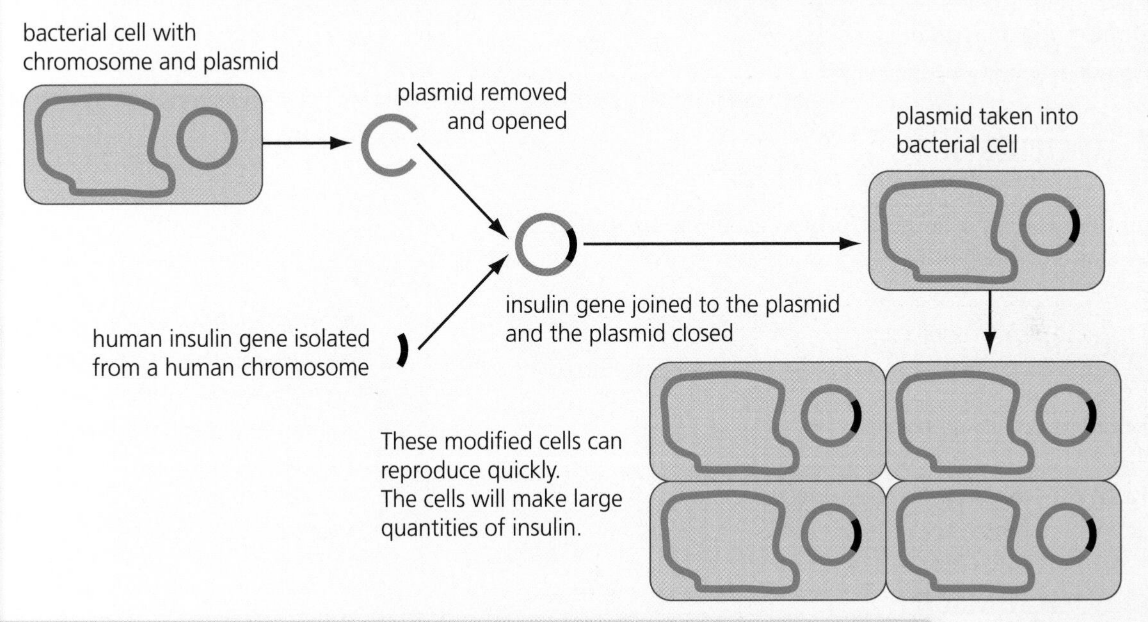

This insulin can be collected, purified and used in the treatment of diabetes.

As people are living longer, there is an increasing incidence of diabetes and therefore an increasing demand for insulin. Traditional sources of insulin were from slaughtered cattle and pigs. The genetically engineered human insulin is easier to obtain and produces fewer side effects in the people who use it as a treatment.

The enzymes in biological detergents can be produced from genetically engineered bacteria. These enzymes can help to clean clothes by digesting the organic component of stains making it easier for water to wash them away. Washing clothes with biological detergents means that lower temperatures can be used. This saves energy by not heating the water as much and is less damaging to delicate fabrics.

The transfer of genes from one species to another is only possible with genetic engineering and wouldn't happen with selective breeding.

Genetic engineering also has the advantage of being much quicker than selective breeding which will take several generations for a new genotype to become established.

Some antibiotics can kill a range of bacterial species but there is no single antibiotic that is effective against all bacteria. In addition, some bacteria can become resistant to antibiotics so to effectively treat all bacterial infections, a wide range of antibiotics are required.

continued

What you should know at Credit level – continued

Enzymes are used in many biotechnological processes. In a batch process, the enzymes remain mixed with the final product and have to be removed. In a **continuous flow process**, the enzymes are **immobilised** by being attached to a surface which allows them to be easily separated from the final product.

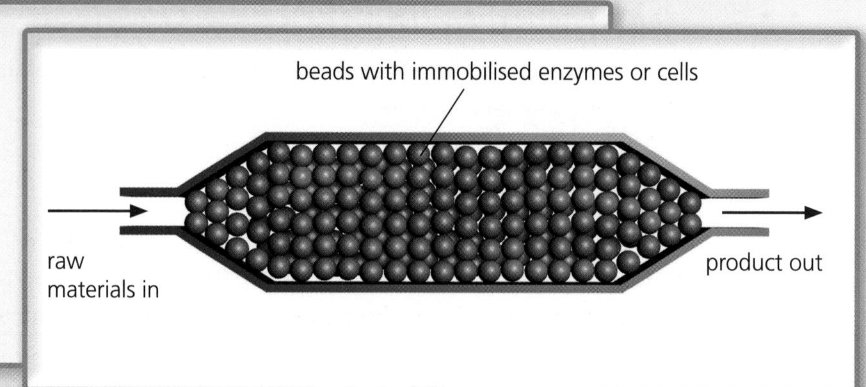

Credit question 1

Explain the need for a range of antibiotics in the treatment of diseases caused by bacteria. (1)

Credit question 1 – Answer

No one antibiotic kills all bacteria.
or *Bacteria may be resistant to some antibiotics.*
or *People can be allergic to some antibiotics.*

Look out for

Remember, it is bacteria and not people that become resistant to antibiotics!

Credit question 2

Immobilised cells are used in some industrial processes. Describe one advantage of using immobilised cells. (1)

Credit question 2 – Answer

Immobilised cells are easily separated from the product.
or *Immobilised cells can be reused.*

The question asks for one advantage. Do not give two answers. If one is wrong you will lose the mark.

Credit question 3

(a) Explain the action of biological detergents. (1)
(b) Explain the economic advantage of using a biological detergent. (1)

Credit question 3 – Answer

(a) *The stains are broken down, or digested, by enzymes.*
(b) *The detergents work at lower temperatures* + *therefore save energy* **or** *cause less damage to clothes.*

(a) The answer must include a reference to enzymes.
(b) You must mention that biological detergents work at lower temperatures and then give a reason why this has an economic advantage.

Answering problem solving questions

The Problem Solving questions of your Standard Grade exam make up half of the total marks and so it is important that you are comfortable with this area of your work.

Problem Solving questions are different from Knowledge and Understanding ones. They do not depend on remembered biological facts but on demonstrating a variety of skills.

The information needed to answer a Problem Solving question is provided in the question itself. You must read the entire question from start to finish to make sure that you find the relevant information.

Problem Solving questions at Credit and General level cover the same skill areas. They differ from each other only in the degree of difficulty. For example, graphs and tables at Credit Level are more complicated and calculations may involve decimal points. The following information on tackling Problem Solving questions, therefore, applies equally well to General and Credit level.

Selecting information

Selecting information involves extracting information from a variety of sources such as; written passages, tables, diagrams, graphs and charts. Remember, the information needed for your answer is there in the question.

Written passages

The examples below are taken from longer passages of text which would usually have five or six questions based on them.

General question 1

"Raptors are completely carnivorous, obtaining all of their required nutrients from their prey. The nutrients which normally come from vegetable matter are often found in the stomachs of their prey."

Explain how raptors can obtain vitamins and minerals found only in plants, even though they are entirely carnivorous. (1)

General question 1 – Answer

These nutrients come from vegetable matter found in the stomachs of their prey.

The question should have given you a clue. It mentions plants, so your answer must refer to plants or at least to vegetable matter. The passage tells us that this vegetable matter is in the "stomachs of their prey", so you have to say exactly where the vegetable matter is found. Simply saying, "They eat their preys' stomachs." or "From their preys' stomachs" is not enough. It does not link the stomachs with the vegetable matter. Notice that you do not need to know what "raptors" are. You are not being tested on knowledge.

Credit question 1

"Many new crop types have been produced. Most of these are modified to be pest, disease or weedkiller resistant, and include wheat, potatoes and onions. Modified crops could become weedkiller resistant 'superweeds'."

Explain why a plant which is modified to be weedkiller resistant could be:

(i) useful to farmers. (1)

(ii) a problem for farmers. (1)

Credit question 1 – Answer

(i) *The crop could be treated with weedkiller to destroy weeds without harming the crop.*

(ii) *The crop itself could become a problem and be difficult to destroy if the farmer wanted to.*

To answer these questions you need to search through the whole passage. You are being asked to explain something. This means that you must show some understanding of the information from the passage. It is not enough to say for part (ii), "They could become superweeds". This adds nothing to what is already in the passage.

Tables

General question 2

Some features of common seaweeds are shown in the table below.

Seaweed	*Colour*	*Shape*	*Bladders*
Bladder wrack	brown	branched	in pairs
Channel wrack	brown	grooved	absent
Cladophora	green	long and thin	absent
Egg wrack	brown	branched	along its length
Sea lettuce	green	flat	absent
Serrated wrack	brown	saw-toothed edges	absent
Spiral wrack	brown	twisted	in pairs

(i) Describe **two** differences between Sea lettuce and Spiral wrack. (1)

(ii) Describe the features which Bladder wrack and Spiral wrack have in common. (1)

You should compare these for part (i).

You are comparing this pair for part (ii).

General question 2 – Answer

(i) 1 *Sea lettuce is green, Spiral wrack is brown.*
2 *Sea lettuce is flat, Spiral wrack is twisted.*
Sea lettuce does not have bladders, Spiral wrack does.

(ii) *They are both brown and have bladders in pairs.*

(i) There are three possible differences so any two of them can be used. It is not enough to say they differ in colour and shape. The details of the differences that can be obtained from the table are needed.

(ii) It is not enough to say that they are the same colour. Again, use the detail that is available.

Credit question 2

The table shows the percentage germination of four crop plants over a range of temperatures.

Temperature (°C)	*Percentage germination of crop plants*			
	Carrots	*Cauliflower*	*Okra*	*Spinach*
0	0	0	0	83
5	48	0	0	96
10	93	58	0	91
15	95	60	74	80
20	96	65	89	52
25	95	53	93	28
30	90	45	88	14
35	74	0	85	0
40	0	0	35	0

Which two crop plants are able to germinate over the widest range of temperatures? (1)

You are often asked to give the range over which something happens. The range means the difference between the lowest and the highest value. It does not mean the highest temperature at which the seeds germinate. Carrots and spinach both germinate over a 30°C range. This is greater than the range for the other crops.

Credit question 2 – Answer

1 *carrots* 2 *spinach*

Charts and graphs

General question 3

The pie chart shows the proportion of injuries resulting from different sports recorded at a sports injury clinic.

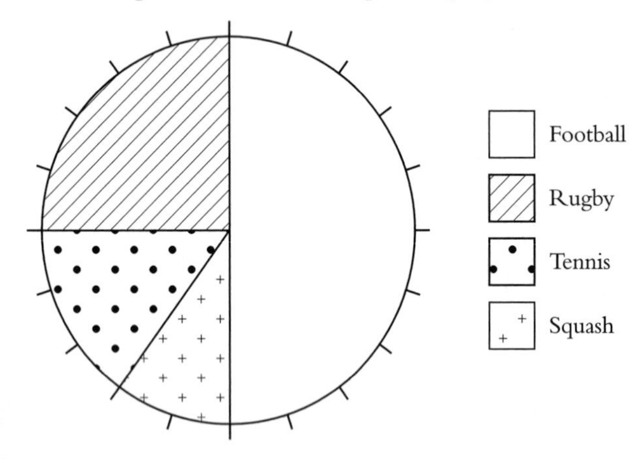

(a) Which sport resulted in 15% of the total injuries? (1)

(b) The number of injuries from playing squash was 32.
How many injuries resulted from playing rugby? (1)

General question 3 – Answer

(a) *Tennis*

(b) *80*

(a) When answering questions on pie charts, you must first work out how much each of the divisions on the chart represents. In this case, there are 20 divisions so each must represent 100% ÷ 20 = 5% of the total. Once you know this, you can see that 15% would be three divisions, which is tennis.

(b) Using the key on the chart, you know that two divisions represent squash. From the question, you know that these two divisions represent 32 injuries. This tells you that each division must be equal to 16 injuries. Rugby takes up 5 divisions – so this must represent 5 × 16 = 80 injuries.

General question 4

The activity of soil organisms was investigated. Some leaves were placed in bags of different mesh sizes and buried in soil for three months.
Each bag was dug up at one month intervals and the percentage decomposition of the leaves recorded. The results are shown on the graph.

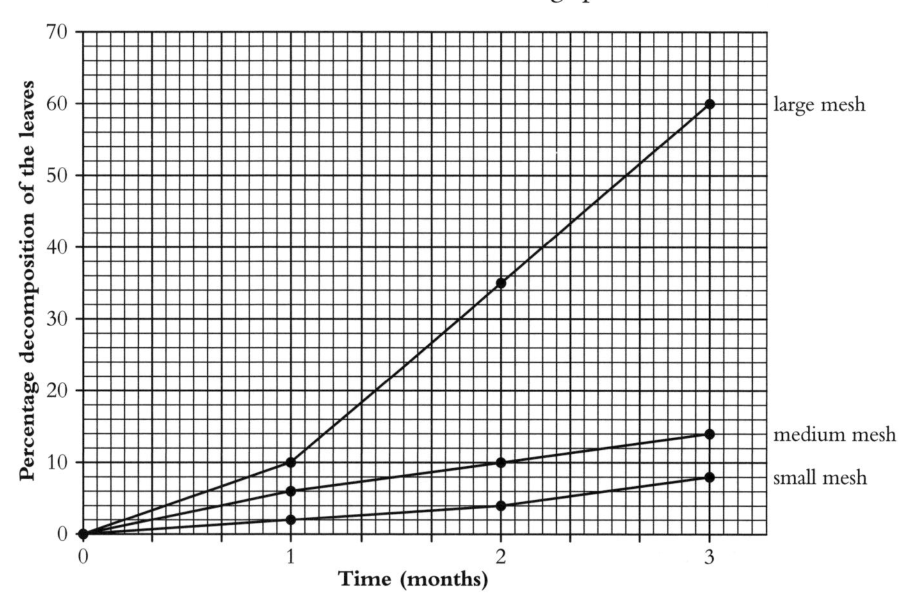

After three months, what percentage of the leaves had decomposed in each bag? (1)

General question 4 – Answer

Large mesh bag ....60.... %
Medium mesh bag ....14.... %
Small mesh bag ....8.... %

When reading values from a graph, you must be sure of the scale used. In this example, each small division of the Percentage decomposition of the leaves represents 2%.

Credit question 3

An investigation was carried out into the effect of light intensity on the distribution of a plant species. At eight different measurement points in a garden, the average light intensity was measured and the percentage ground cover of the plant was recorded.
The results are shown below.

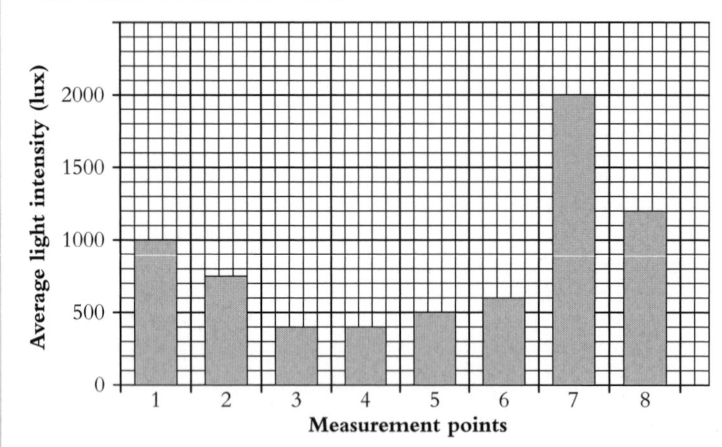

Measurement points	*Ground cover of the plant (%)*
1	85
2	65
3	20
4	20
5	30
6	35
7	100
8	90

(i) What is the light intensity in the garden where the ground cover of the plants was 100%? (1)

(ii) What was the percentage ground cover of the plants when the light intensity was 750 lux? (1)

Credit question 3 – Answer

(i) *2000* lux
(ii) *65* %

Sometimes you are asked to select information from more than one source. To answer each of these questions, you need to refer to both the table and the chart.
(i) The table tells you that the ground cover is 100% at measurement point 7. The chart shows that at measurement point 7, the light intensity is 2000 lux.
(ii) This time you need to consult the chart first. From the chart, you can see that the 750 lux value is present at only measurement point 2. The table shows that measurement point 2 has 65% ground cover.

Presenting information

Presenting Information involves taking information or data given in the question and presenting it in a different form. This could be completing a table of results, completing a graph, chart or key.

Tables

Credit question 4

The table shows the percentage germination of four crop plants over a range of temperatures.

Temperature (°C)	*Percentage germination of crop plants*			
	Carrots	*Cauliflower*	*Okra*	*Spinach*
0	0	0	0	83
5	48	0	0	96
10	93	58	0	91
15	95	60	74	80
20	96	65	89	52
25	95	53	93	28
30	90	45	88	14
35	74	0	85	0
40	0	0	35	0

Complete the table below by adding the correct heading, units and values to show the optimum germination temperature for each of the crops. (2)

Credit question 4 – Answer

Crop plant	Optimum germination temperature (°C)
Carrots	20
Cauliflower	20
Okra	25
Spinach	5

The first point when completing a table is to make sure the column headings are correct. Here you are asked for the optimum germination temperature of each crop – so that is what you put as the column heading. Do not alter it or try to put it into your own words! Remember to include the units when appropriate.
To complete this table you need to know that 'optimum' means the point when something works best – in this case the temperature that is best for germination for each type of seed.

Keys

Usually at General Level you would be asked to complete a branched key while at Credit Level a paired statement key will normally be used. The information needed to complete any key may be presented in a table or given as diagrams.

If it is from a table, make sure the features you use in the key quote exactly the information from the table. If it is from diagrams, make sure the features you decide to use can be clearly seen on the diagrams.

General question 5

Some features of six members of the buttercup family are shown in the table below.

Plant name	*Leaves*	*Runners*	*Stem*
Greater spearwort	toothed	present	hairy
Meadow buttercup	lobed	absent	hairy
Lesser celandine	heart-shaped	absent	hairless
Creeping buttercup	lobed	present	hairy
Lesser spearwort	toothed	absent	hairless
Celery-leaved buttercup	lobed	absent	hairless

Use the information in the table to complete the key below. Write the correct feature on each dotted line and the correct names in the empty boxes. (3)

Two points to look out for here. Firstly you must make sure that you know what you are supposed to write on the dotted lines and what goes in the boxes. In this question, you are told that you must write the feature on the dotted line and the name of the plant in the boxes. Even if you had not been given that information, you could have worked it out by looking at the diagram. All of the plant names which have been provided are in the boxes while the dotted lines are lined up with the other features which describe the plants.

continued on page 97

General question 5 – Answer

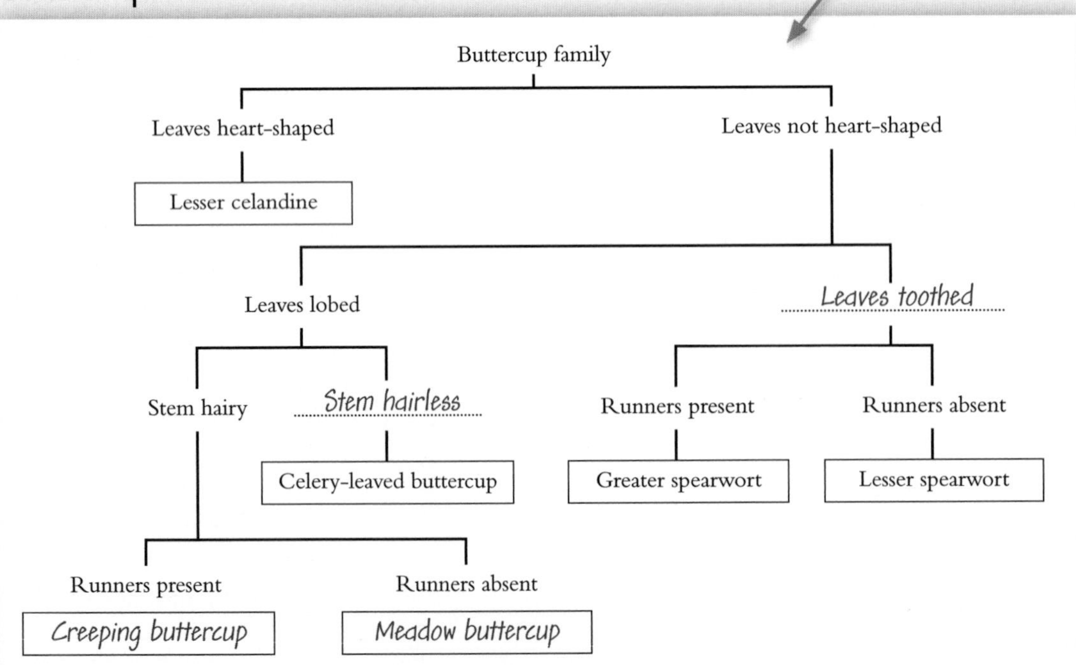

General question 5 – Advice continued...

The next point is to notice that when you write in the feature, you must provide the full description. It is not sufficient to write, 'toothed' or 'hairless'. You must write, 'leaves toothed' or 'stem hairless'.

Completing the key is simply a matter of matching the descriptions with the table. The right-side of the key starts by looking at leaf shape. From the table, you can see that both of the spearworts have toothed leaves so that description must go on the first dotted line. You can also see that the celery-leaved buttercup has lobed leaves but a hairless stem so you would write that on the second dotted line. Similarly, by working through the table, you can identify creeping buttercup and meadow buttercup. Again you must give the full names. You would not get a mark if you wrote only, 'creeping' or 'meadow'.

Credit question 5

The table below describes the features of the fluid which lead to diagnosis of several joint abnormalities.

		Feature of synovial fluid		
		Viscosity	*Cloudiness*	*Colour*
Diagnosis	*Normal*	high	zero	light yellow
	Inflammation	low	slight	dark yellow
	Infection	low	high	dark yellow
	Blood leakage	intermediate	high	pink

Use the information from the table to complete the paired statement key to identify the diagnoses. (2)

Credit question 5 – Answer

1	Fluid pink	Blood leakage
	Fluid not pink	go to 2
2	Low viscosity	*go to 3*
	High viscosity	*Normal*
3	*High cloudiness*	Infection
	Slight cloudiness	*Inflammation*

From Statement 1 you know that all the fluids described in 2 are a colour other than pink. Looking at the table, you can see that there are three which fit this description but only one of these has high viscosity. This allows you to identify it as "Normal". To distinguish between the remaining two, you need to put, 'go to 3' in the box opposite Low viscosity. These remaining fluids are both dark yellow so we need to use 'Cloudiness' to separate them. From the table, Infection results in High cloudiness so we can write that in the first box in statement 3. This leaves only 'Slight cloudiness' and 'Inflammation' for the remaining boxes. Notice you must write 'Slight cloudiness' not just 'Slight'.

Charts and graphs

You may be asked to present a set of results in a variety of ways including pie charts, bar graphs and line graphs.

If you make a mistake with a chart or graph, there will always be an additional chart or graph found at the end of the paper. It is very important that you complete the whole graph on the spare copy. This includes putting in the labels and scale and not just drawing the graph. Your mark will be based on only one of the graphs. You cannot get a mark based on the combined answers for the two.

Take care when drawing lines on any graph or chart. Use a ruler and make sure the line is where it should be.

Pie charts

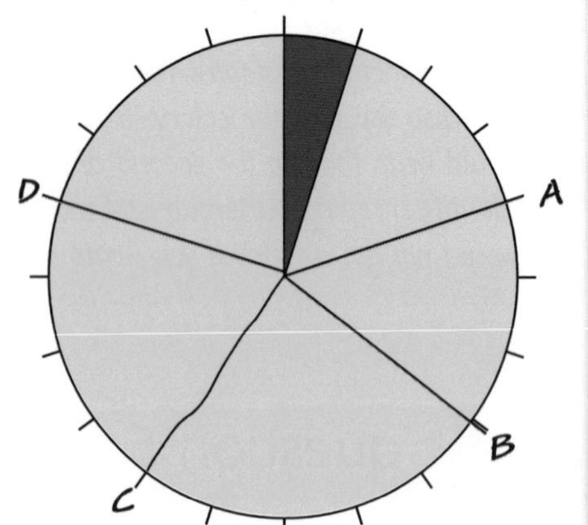

If you had completed this pie chart, only one of the lines would be correctly drawn.

Line A is fine.

Line B would not pass. It is important that there is no gap between the line and the segment marker on the edge of the circle. In this case there is a clear gap so no marks.

Line C has been hand drawn and so the area contained in the segment of the circle cannot be accurate.

Line D looks as if it is correctly drawn but, although it is perfect at the edge, it has missed the centre of the circle and no marks would be given.

To find out what percentage each division represents, simply divide 100 by the total number of divisions marked out round the circle. Your answer is the percent value of each segment.

If you are asked to complete a pie chart based on numbers rather than percentages, you will always be given some information about at least one of the segments. This will enable you to work out the value of the other segments.

General question 6

The composition of food made from a single celled micro-organism is shown below.

Component	*Percentage*
protein	45
fat	10
minerals	5
fibre	5
other nutrients	35

Use the information in the table to complete the pie chart shown below. (2)

General question 6 – Answer

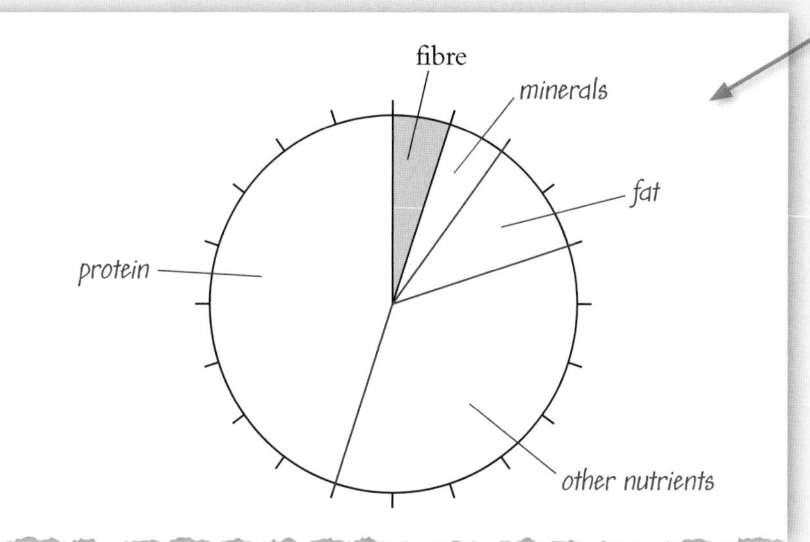

Since the segment for fibre has already been drawn for you, it is clear that each division in the chart must represent 5%. Using this information, you can complete the chart. You can check your answer by comparing your chart with the table. The biggest number in the table will be the biggest segment of the chart, and so on. Notice that not only must the lines be drawn correctly for one mark but each segment must be correctly labelled using the exact description from the table in order to get the second mark. For example you must write, 'other nutrients', not simply, 'others' or 'nutrients'. The segments must be labelled with the names of what they represent – not with the percentages from the table.

Bar charts, bar graphs and line graphs

Remember, if you are asked to draw a line graph, draw a line graph – not a bar chart. A line graph is not a series of lines standing up on end. It is a line joining a series of plotted points.

Curves of best fit are not accepted in Standard Grade Biology – you must join the points using a ruler.

Questions involving the completion of graphs usually ask you to choose a scale and label an axis, as well as plotting the points and drawing the line or drawing the bars.

When choosing a scale, select one that will make it easy for you to plot the points. If there are any points already plotted on the graph, this will allow you to work out what the expected scale should be. The scale you use must go past the last value to be plotted. If the highest plot point represents 37 units, your scale should go up to at least 40. In addition, you have to choose a scale which will allow you to use at least half of the graph paper. The graph paper which is in your examination booklet will usually allow you to use the whole graph paper at a sensible scale.

When you label the axis, use the exact words and units which appear in the table of results that you are using.

Plot only the values you have been given. Do not draw the line back to zero on the x axis unless you have been given a value for this point.

continued

Bar charts, bar graphs and line graphs – continued

The following question illustrates the important points regarding drawing graphs.

Using the information in the table, draw a line graph to show the relationship between temperature and the rate of photosynthesis.

Temperature (°C)	5	10	15	25	30	40
Rate of photosynthesis (bubbles/minute)	4	8	12	20	26	32

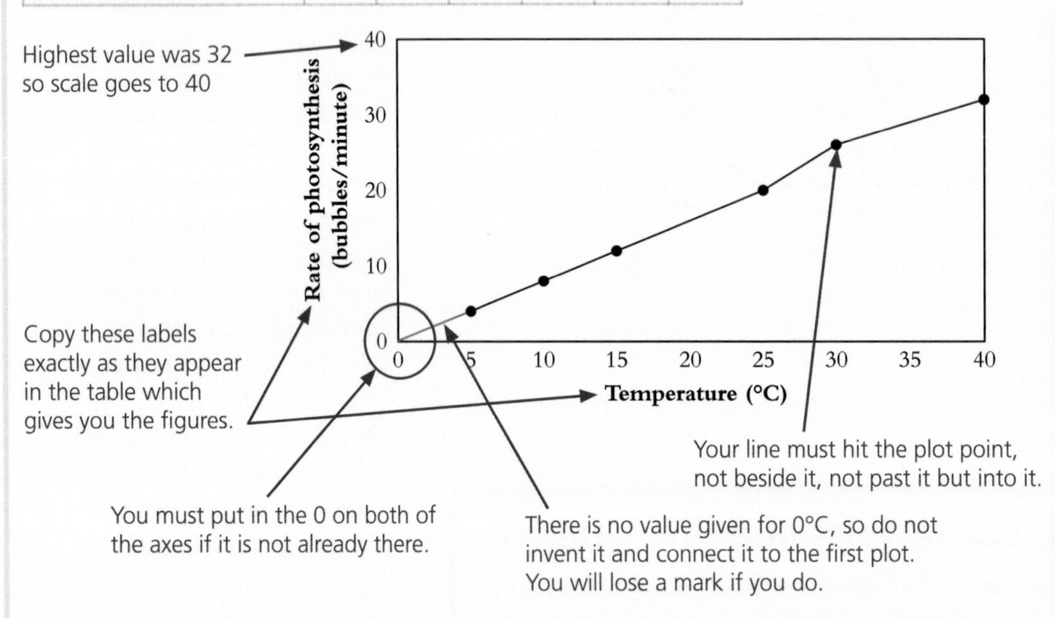

This graph shows that when you put in your scale on the axes, it must go up in regular divisions, even where there are no values to be plotted. In the example above, you are given temperature values of 5°C, 10°C, 15°C, 25°C, 30°C and 40°C, with no data for 20°C and 35°C, but you must put in 20°C and 35°C on the scale along the x-axis.

When completing bar charts, take great care with the tops of the bars. In the example below, only one of the six is correctly drawn to indicate the marked line.

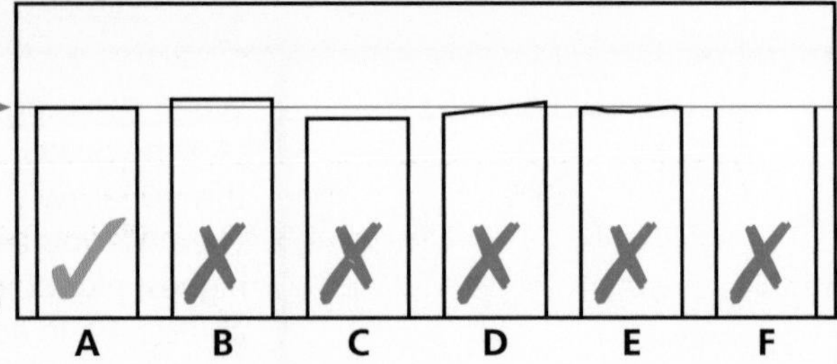

Bar A is correctly drawn. The top of the bar is on the line with no gaps between the top of the bar and the line.

Bars B and C both show gaps between the tops of the bars and the line. If there is a gap like this you will lose the mark.

Bar D has part of the top above the line and part below. This leaves the marker wondering which you intended.

The top of bar E has not been drawn with a ruler and loses the mark.

Bar F does not have a top and so this bar would again lose the mark. If you do shade in a bar, make sure that the shading does not go above the line you have drawn at the top of the bar.

Credit question 6

The table shows the occurrence of chromosome mutations in Drosophila fruit flies when exposed to different doses of radiation.

Dosage of X-rays (millisieverts)	*Chromosome mutations* (%)
1000	1·0
2000	1·9
2500	2·6
3000	3·1
4000	4·2
4500	4·6
5000	5·3

Use this information to draw a line graph. (3)

Credit question 6 – Answer

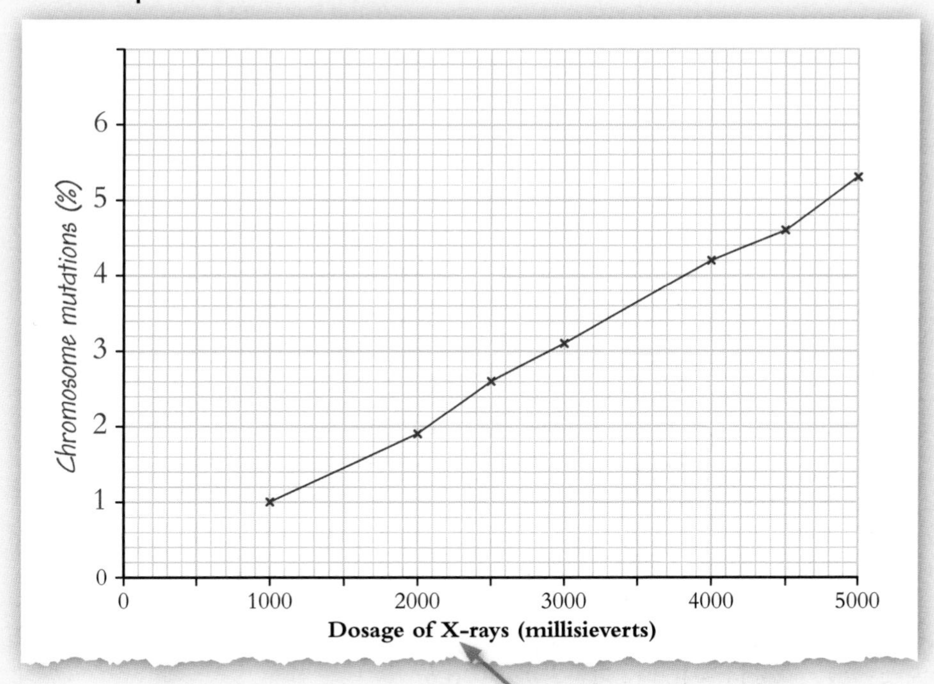

If you are asked to label the y-axis, you are expected to copy the label from the results table onto the appropriate axis. Do not put, 'y'. This is not a label. In this question, the 0 was already provided on the y- (vertical) axis. Remember that if it had not been there, you would have had to add it. Many marks are lost by not carrying out this simple step!
Plotting these points needs some care because the decimal points which are odd numbers will be mid-way between the gridlines. Where this happens, you should ensure that there is a gap above and below the plot point so that there can be no doubt that you meant it to be in the space and not on a line.

General question 7

The table shows the results of treating an infection in cows with various antibiotics.

Antibiotic treatment	*Number of cows treated*	*Number of cows cured*	*Percentage of cows cured*
no antibiotic	2011	1450	72
Amoxicillin	56	48	86
Cephapirin	18	16	89
Cloxacillin	33	25	76
Erythromycin	8	6	75
Penicillin	25	17	68

Use the information in the table to complete the bar chart to show the percentage of cows cured by:

1 labelling the vertical axis
2 adding a scale to the vertical axis
3 completing the bars. (3)

The information you need to add to complete the bar chart successfully is clearly stated in the question at General level. At Credit level you would be expected to know that you should label the axis and add a suitable scale. The 'None' column has been completed for you.
The table tells you that 72% of cows were cured in this group so this lets you work out what each division of the graph represents; in this case 2%.

General question 7 – Answer

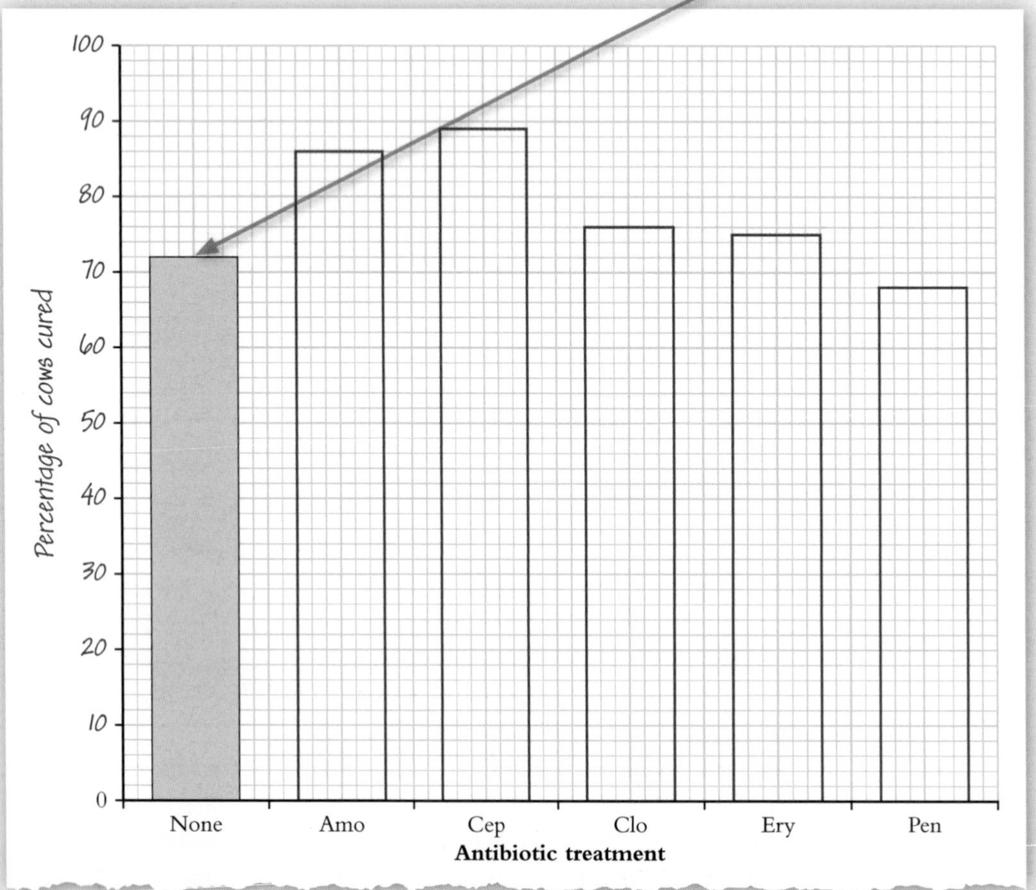

Calculations

It is not the purpose of this topic to teach you basic arithmetic but rather to put it in the context of biology. If you are good at arithmetic and can carry out the various calculations mentioned here, stay with what you have learned in school rather than trying to learn a completely different method. If you feel that you are having difficulty with the calculations in your tests and homework, you might want to have a read through the following pages to see if these methods make more sense to you.

All General level calculations will give answers in whole numbers. If your answer is not a whole number then it is wrong. The calculations at Credit level are similar to those at General level but are more difficult and whole numbers are not always used.

You often have to get information from a table or graph before you can do the calculation.

Look at the figures in the question. If they have been rounded up to whole numbers or to one decimal point, then you should do the same.

Averages

The average of a set of numbers is the total of these numbers divided by how many numbers there are in the set.

General question 8

A population survey of barnacles on a rocky shore was carried out using quadrats.
The results are shown in the table.

Quadrat number	1	2	3	4	5	6	7	8	9	10
Number of barnacles	52	51	37	40	40	23	15	17	15	10

Calculate the average number of barnacles per quadrat. (1)

General question 8 – Answer

Average number ......30......

The answer at General level will always be a whole number.
An easy way of telling if your answer is wrong is that a correct answer in averages will always be between the highest and lowest figure that you have added up.

The sum = 52 + 51 + 37 + 40 + 40 + 23 + 15 + 17 + 15 + 10 = 300
The average = 300 ÷ 10 = 30

General question 9

The effect of practice on the reaction times of three volunteers was investigated.
A buzzer was sounded and the time taken to stop a clock was measured.
Each volunteer was tested 10 times.
The results are shown in the table.

	Reaction time (milliseconds)									
Volunteer \ *Attempt*	1	2	3	4	5	6	7	8	9	10
A	256	250	210	207	201	192	187	164	162	154
B	234	227	218	201	200	185	179	161	153	147
C	218	200	195	192	186	178	160	149	136	131

The average reaction time of the three volunteers' first attempts was 236 milliseconds.
Calculate the average reaction time of their final attempts. (1)

General question 9 – Answer

Average ......*144*...... milliseconds

To answer this question correctly, you must make sure that you understand what is being asked. "Their final attempts..." does not mean the last two or three attempts for each volunteer. It means the average of the final attempt of all three, that is, Attempt 10.
Total of the times for Attempt 10 = 154 + 147 + 131 = 432
Average = 432 ÷ 3 = 144

Ratios

A ratio is used to compare two quantities. When answering questions about ratios it is important to:

- use the same units for all quantities
- keep the order of the quantities the same as it appears in the question.

General question 10

In an investigation into the inheritance of colour in onions, two red onions were crossed to produce 36 red and 9 white ones.
Calculate the simple whole number ratio of red onions to white onions in the offspring. (1)

General question 10 – Answer

It is very important that you give the answers in the correct order. 4:1 is not the same as 1:4. Always read the question carefully.
A good place to start when working out ratios is to try to divide all the figures by the smallest. In the General paper, this should result in a ratio where the figure for one side is 1. For example, in the question above, 9 will divide into 36 without a remainder. In the Credit paper, things are not always so simple but try it anyway. If it works then you do not have to do any more with your answer. If the smaller figure will not divide into the larger without a remainder or without producing a decimal point in the answer, stop and try another approach.
Here are some suggestions:

- *If all the figures end in a 0, they will all divide by 10. If they end in a 0 or 5, they will all divide by 5.*
- *If the figures are all even numbers, they will divide by 2.*
- *If you get to the stage where you have divided by 10, 5 or 2 and neither number you are left with is a 1, try dividing both sides by 3 just to see if you can get the numbers any smaller but it might be that this is as far as the ratio goes while still remaining whole numbers. Remember, the question will always ask for a simple whole number ratio.*

If you are dealing with three numbers and one of them is a 0, it will stay 0 in the final ratio. You simply carry on with the other two to work out the ratio between them.

Credit question 7

In an investigation, the kidneys of an adult male were found to filter 1254 cm^3 of blood per minute. This produced 114 cm^3 of filtrate per minute and 1·2 cm^3 of urine per minute.
Express these volumes as a simple whole number ratio. (1)

Credit question 7 – Answer

*At first sight this looks like a difficult question but if you remember the first suggestion above, you will find that all of the numbers will divide by the smallest, 1·2, resulting in a whole number ratio. The fact that one of the numbers in your answer is very high compared to the smallest is not important. Once you have reduced one of the numbers to 1, that is as far as you can go. It helps to get them in the right order if you write them down in the correct order in the **Space for calculation** in the paper before you start trying to work out the answer.*
Sometimes you are asked to use a given ratio to calculate the actual number of one group as in the following question.

Credit question 8

The table shows the number of people with each blood group in a population of 1500.

Blood group	Number of people
A	610
B	143
O	675
AB	72

In the population, the ratio of males to females with blood group AB is 5:3.
How many males would have blood group AB? (1)

Credit question 8 – Answer

45

You are told that there are 72 people with blood group AB and that the ratio of males to females is 5:3.
It is this ratio that allows you to work out the answer. Add both sides of it together:
5 + 3 = 8
Now divide that answer into the total number of people with AB, in this case 72:
72 ÷ 8 = 9
Multiply this by the ratio of males to get your answer,
9 × 5 = 45.

Percentages

Percentage calculations are guaranteed to appear in your Biology paper so you should try to make sure that you can tackle them successfully. Unfortunately there are probably as many ways to teach percentage calculations as there are teachers so if you feel that you can do them, don't confuse yourself with the details of the following examples.

If you understand that 'percent' means 'for every hundred' then you are on your way to being able to do the problems.

Here are the main ways in which you would be expected to use percentage calculations:

1 To express a number as a percentage.
Example – A class of 30 pupils contained 18 boys. Calculate the percentage of boys in the class.
This means that you must express 18 as a percentage of 30.

Rule 1:		
	Take the number you want to express as a percentage	18
	Multiply it by 100	= 1800
	Divide the answer by the total (30)	= 1800 ÷ 30
		= 60%

2 To convert a percentage value into a number.
Example – 16% of the trees in an area of woodland were oak trees. There were 525 trees in total. How many oak trees were there?
This means that you calculate 16% of 525.

Rule 2:		
	Take the percentage you want to convert	16%
	Multiply it by the total (525)	= 8400
	Divide the result by 100	= 8400 ÷ 100
		= 84

continued

Percentages – continued

3 To calculate a percentage change.
Example – What was the % change between a resting pulse rate of 70 and a rate when walking of 98?
This means that you must first find the difference between 70 and 98 and then calculate that number as a percentage of the starting value.

Rule 3: Find the difference between 70 and 98 = 28
Multiply the answer by 100 = 28 × 100
= 2800
Divide the difference by the original number (70) = 40%

If you have been following this, you will notice two things: every calculation involves 100 and you multiply or divide *only once* in each calculation. So if the rule tells you to make a number into a percentage you should multiply by 100, then you know that you will have to divide by the remaining number.

General question 11

The eye colours of 160 school pupils are shown in the table below.

Eye colour	*Number of school pupils*
brown	80
green	24
blue	48
grey	8

What percentage of the school pupils had green eyes? (1)

General question 11 – Answer

15%

Following Rule 1 above, to make a number into a percentage, multiply by 100.
So 24 × 100 = 2400
Divide the answer by the total in the group
2400 ÷ 160 = 15%

General question 12

In an investigation into breathing rates, a pupil, has his number of breaths per minute first recorded when standing still then when exercising at different levels.
The procedure was repeated three times and the results are shown in the table below.

	Breathing rate (breaths per minute)			
Exercise	*1st trial*	*2nd trial*	*3rd trial*	*Average*
standing still	16	15	17	16
walking	19	17	18	18
jogging	27	25	29	27
running quickly	33	31	32	32

Calculate the percentage change in the average breathing rate when running quickly, compared to standing still. (1)

General question 12 – Answer

100%

Use rule 3 to find a percentage change.
The breathing rate goes up from 16 to 32 so the difference is
32 – 16 = 16.
Multiplying the difference by 100 gives
16 × 100 = 1600
Dividing this by the original to give the percentage is
1600 ÷ 16 = 100%

Credit question 9

The table shows the percentage germination of spinach seeds over a range of temperatures.

Temperature (°C)	% *Germination*
0	83
5	96
10	91
15	80
20	52
25	28
30	14
35	0
40	0

What is the minimum number of spinach seeds which should be sown at 15°C in order to produce 1000 seedlings? (1)

General question 12 – Answer

1250 seeds

There are various ways to tackle this problem. Here is one way:

- At 15°C, 80 seedlings are produced for every 100 seeds sown. However, we need 1000 seedlings.
- This is 12·5 times more than 80. (1000 ÷ 80 = 12·5). So if we want 12·5 times more seedlings, we will need to sow 12.5 times more seeds. 12·5 times more than 100 is 1250 (12·5 × 100).

Or you could try this:

- In order to get 80 seedlings, you would have to sow 100 seeds.
- Therefore, if you wanted to get one seedling (unlikely but possible!) you would need to sow on average 100 ÷ 80 seeds = 1·25.
- However, you are not trying to get 1 seedling but 1000 seedlings so you would need to sow 1000 times as many seeds = 1·25 × 1000 = 1250 seeds.

This question goes beyond simple percentages but is included in case you meet something similar in future.

Mixed skills

It is unusual for Problem Solving questions to test only one area of problem solving. A graph question, for example, might ask you to complete the graph, read information from the graph, draw conclusions and make a prediction. Similarly a question based on calculations can test a variety of skills. Here is an example of a typical Problem Solving question. Notice that it covers three different skills.

General question 13

The table below gives information about the harvest of softwood timber over a 3 year period.

Country	*Timber harvested each year* (m^3)		
	1999	*2000*	*2001*
Scotland	2292	2496	2883
England	699	881	913
Wales	385	463	663
Total	3376	3840	4459

(a) Complete the totals to show the totals harvested in 1999 and in 2001. (1)

(b) What was the average annual timber production in England for the years shown? (1)

(c) What percentage of the total timber harvested in 2000 was produced in Scotland? (1)

(a) Some students are confused when there is no line after the Space for calculation on which they can write their answer. The question asks you to complete the totals. In other words, you are expected to write your answers in the spaces in the table. You do not have to add the units as they are already given in the table headings.

General question 13 – Answer

(b) 831 m^3

(c) 65%

(b) The only way in which you could lose a mark here is through carelessness. Make sure you select the correct line, add up the figures correctly and divide by three. You must add the units (in this case m^3) if they do not already appear on the answer line.

(c) In this question again, it is important that you select information correctly.

- *You need to look at the correct year and you will find that the total for 2000 was 3840 m^3.*
- *Scotland's production was 2496 m^3.*
- *Since you are making a number into a percentage, you need to multiply by 100.*
- *This gives 2496 × 100 = 249600.*
- *Remember you use multiply and divide once only in these calculations and since you have already multiplied by 100, you must have to divide by 3840 to get the answer. 249600 ÷ 3840 = 65%*

Sometimes your calculations will be based on extracting information presented in a graph. The same basic rules apply.

Credit question 10

The graph shows the number of kidney transplants carried out and the number of patients waiting for a transplant in the UK between 1996 and 2005.

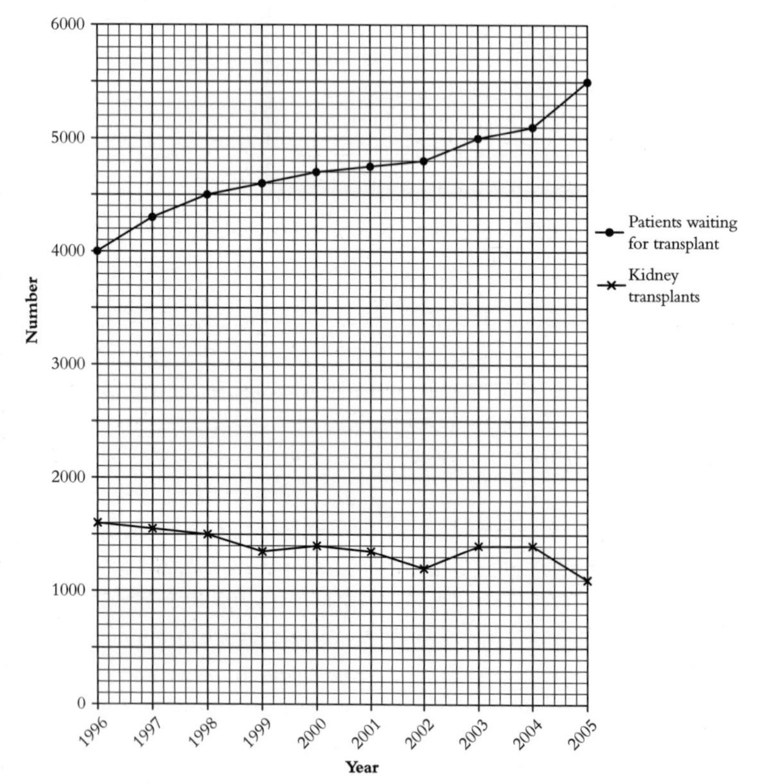

Calculate the average yearly increase in the number of patients waiting for a transplant from 2000 to 2005.

Credit question 10 – Answer

160

A common mistake in this type of question is to try to add up the increase for all the individual years – not only does this take up far too much time, it is also wrong!
You are asked for the average yearly increase, so all you need to do is to find the total increase and divide it by the number of years.

Total increase is the number waiting in 2005 minus the number waiting in 2000: = 5500 – 4700 = 800
Number of years involved: = 2005 – 2000 = 5
Average yearly increase is = 800 ÷ 5 = 160

Conclusions and experimental procedures

Questions on experimental procedures test your ability to:

- question the validity of an investigation
- suggest improvements if you see an error in the set up
- draw conclusions from the results of the investigation
- make predictions about the outcome of an investigation.

Experimental based questions

Valid or **reliable**? These are two terms which crop up a lot in questions about investigations.

If an investigation is valid, it means that the result was caused by the factor you were testing, light intensity, pH, temperature, etc. and could not have been caused by anything else. The way you make an investigation valid is by setting up a **control**. A control is an experiment which is identical to the original except for the factor you are investigating. In this way you can compare the two to make sure that the factor you are investigating really is causing the change. If the question asks, "What is the function of a control?" then you can answer it in general terms. "To show that the result must be caused by the factor being investigated" is a good answer. If the question asks, "What is the function of a control in this experiment?", then you must answer it in terms of the actual experiment, for example: "To show that it is the light intensity which is affecting plant growth."

The results of an experiment are more reliable if you can reduce the effect of an atypical (fluke) result. The way to do this is to repeat the experiment many times or to increase the number of seeds, beetles and so on that you use in your experiment. You can never get 100% reliability so if asked to comment on how repeating an experiment will affect the reliability, you can only say it will, 'increase the reliability' or 'it will make it more reliable'. You cannot say, 'it will make it reliable'.

Drawing a valid conclusion

Drawing a conclusion involves thinking about the information that is given and deciding what we can learn from that information. The information will often be in the form of a table of results or results displayed in a graph or chart.

General question 14

An investigation was carried out into the effect of lemonade on teeth. Three teeth were placed in a tube containing 20 cm^3 of lemonade. Another three teeth were placed in 20 cm^3 of water. All other factors were kept the same. After 24 hours the teeth in the lemonade had lost 9% of their mass, whilst the teeth in water had not lost any mass.

(a) The tube with water instead of lemonade was a control. What is the purpose of a control? (1)

(b) The teeth were sterilised before carrying out the investigation. Explain why this was necessary. (1)

(c) Give two factors, not mentioned already, which would need to be kept constant for the investigation to be valid. (2)

(d) What conclusion can be drawn from this result? (1)

General question 14 – Answer

(a) *A control shows that the result must be due to the factor being investigated and not to any other factor.*

(b) *To ensure that bacteria on the teeth were unable to affect the result.*

(c) Any two from:
temperature *hardness of teeth* *surface area of teeth*
mass of teeth *concentration of lemonade*

(d) *Lemonade dissolves teeth*

Look out for

Does the question require a general definition of a control, or does it ask you for an answer that is particular to this question?

(a) This is a good definition for a control – you might want to memorise it. Notice that this question asks for a general answer. The question could have asked for the purpose of the control in this particular investigation. Then you would answer using the actual variables involved. ("the control in this investigation shows that the effect on the teeth was caused by the lemonade and by no other factor")

(b) Sterilisation means destroying all living organisms. Living bacteria cause teeth to decay and so would affect the result.

(c) There are many acceptable answers, but just check that the ones you choose have not been mentioned already in the question.

(d) The information is that teeth in lemonade lost mass but teeth in water did not. Where has the material from the teeth gone? It must have dissolved. That gives us our conclusion.

General question 15

In an experiment, a suspension of the bacterium *Escherichia coli* (E. coli) was spread evenly over nutrient agar in petri dish 1. In the same way, the bacterium *Staphylococcus albus* (S. albus) was spread over nutrient agar in petri dish 2.

A paper disc containing the antibiotic streptomycin was placed in the centre of each petri dish.

The diagrams show the appearance of the two dishes after 48 hours.

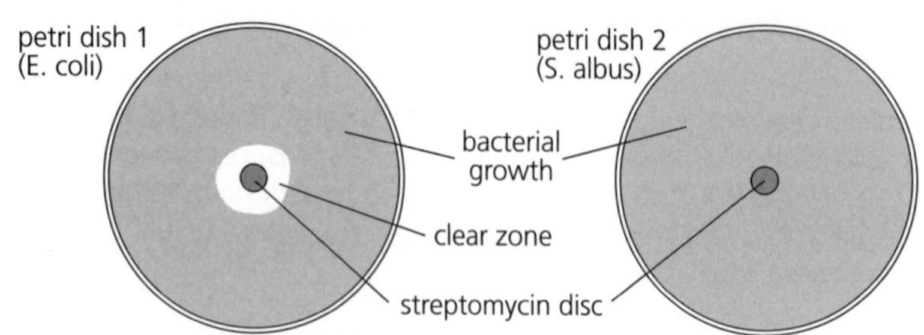

(a) What conclusion could be drawn about the effect of streptomycin on the two bacteria? (1)

(b) Describe the control which should be set up to ensure that this conclusion is valid. (1)

General question 15 – Answer

(a) *E. coli* *killed by streptomycin*
S. albus *unaffected by streptomycin*
(b) *Set up another two petri dishes exactly the same as those in the diagram, but with paper discs containing no antibiotic.*

(a) The information shows a clear zone next to the streptomycin in E. coli, but bacterial growth right up to the streptomycin in S. albus. You know that antibiotics kill bacteria so the conclusion is that streptomycin kills one but not the other.
(b) When asked to describe a control, always start by saying 'Another set-up, identical in every way except ...' If you only said 'no antibiotic' in this case you would not get the mark – you need to say that everything else is the same, and to say what has to be changed.

General question 16

Catalase enzyme releases oxygen from hydrogen peroxide. Different tissues were tested for catalase activity by adding equal masses to hydrogen peroxide at pH 7.
The height of the foam gave a measure of the volume of oxygen released.

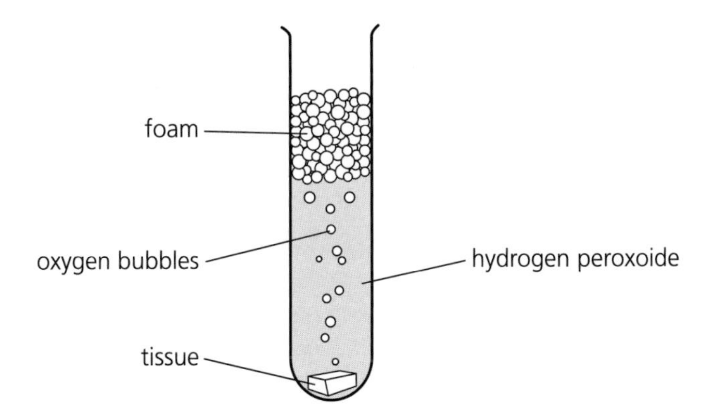

Beef, fish and chicken tissues produced greater volumes of oxygen than carrot, apple or potato.
(a) Suggest a hypothesis which could explain this fact. (1)
(b) Give one variable, other than pH, which must be kept constant in this investigation. (1)
(c) Describe a suitable control for this investigation. (1)

General question 16 – Answer

(a) *Animal tissues have more catalase than plant tissues.*
(b) Any of:
temperature *mass of tissue*
surface area of tissue *concentration of hydrogen peroxide*
(c) *The control must be identical to the original but with something non-living instead of the tissue (e.g. glass, Plasticene, etc.)*

(a) A ***hypothesis*** *is a theory which tries to account for the results of an experiment. The question tells us that beef, fish and chicken produced more oxygen and that oxygen was produced by catalase activity. Beef, fish and chicken are all animal tissues and this gives us our hypothesis.*
(b) Make sure you use an example that applies to this particular investigation.
(c) Notice that you must keep the overall volume of the apparatus the same, so you need something to replace the missing volume of the tissue.

Credit question 11

In an investigation into behaviour, five leeches were placed in water in a shallow rectangular dish as shown in the diagram.

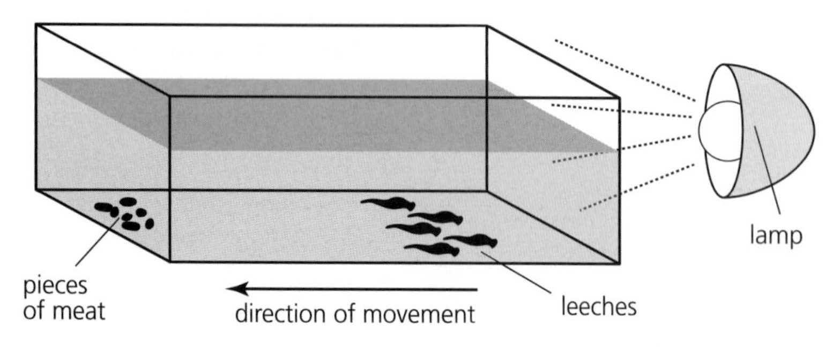

(a) During the investigation the leeches moved in the direction shown. Give two possible explanations for this response. (2)

(b) Suggest one change which should be made to the set up of the investigation so that only one valid conclusion could be drawn from the leeches' response, assuming the direction of movement stays the same. (1)

(c) Why were 5 leeches used rather than one? (1)

Credit question 11 – Answer

(a) 1 *the leeches move towards the smell of food*

2 *the leeches move away from the light*

(b) *carry out the experiment in darkness*

(c) *This increases the reliability of the results by minimising the effect of any atypical results.*

(a) Don't be intimidated! You don't need any knowledge of leeches to answer this question. Other than the leeches there are only two things in the procedure – the light and the food. Clearly the behaviour must involve one or the other.

(b) You need to eliminate the effect of one variable to make the procedure valid. So you simply switch off the light to see only the effect of the food, or you don't put the food in, and observe the effect of the light.

(c) One of the most frequently asked PS questions. Rather than just say 'increases reliability', it is safer to add the bit about minimising the effect of atypical results. Remember, results can never be 100% reliable, so do not say 'It makes the results reliable' – you won't get the mark.

Part of the point of questions in the context of experimental situations is to give you the opportunity to show that you know the basic principles and can apply them in new contexts. So do not let the details panic you. It does not matter if the chemical reactions, animals, plants or apparatus are completely weird - all you need to know is what is required for valid measurements and reliable results.

Credit question 12

An investigation on the effects of the presence of plants on soil erosion was carried out.
The diagram below shows the apparatus used.

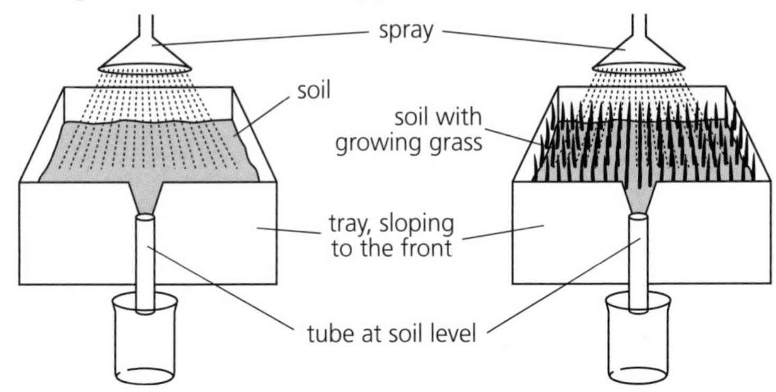

Tray A contained soil only. Tray B contained the same type of soil but with grass growing in it. Water from a spray was allowed to fall onto each tray. Any water running off was collected and examined for soil washed from the tray.

(a) (i) Identify the variable which was altered in the investigation. (1)

(ii) Describe two variables which must be kept constant in both sets of apparatus during the investigation. (1)

(b) The mass of soil washed from each tray was measured by filtering the soil from the collected water. The soil was dried in an oven set at 90°C and then weighed.

(i) Identify a possible source of error in the drying and weighing procedure for measuring the mass of soil washed from each tray. (1)

(ii) Explain how this error could be minimised. (1)

(c) The results showed that tray A had 14·7 g of soil washed from it whilst Tray B lost only 2·6 g of soil. What conclusion can be drawn from these results? (1)

Credit question 12 – Answer

(a) (i) *the presence or absence of grass growing in the soil*

(ii) Two from:

The flow rate of the sprays *The angle of slope of the trays*

The height of the spray above the tray

(b) (i) *The soil might not be completely dry so the weight of some water could be included in the measurements.*

(ii) *Continue drying until the weight stays the same.*

(c) *The presence of grass growing on soil reduces soil erosion by water.*

a) (i) The only difference between the two set-ups is the grass – so that is the variable that was altered.
(ii) The possible answers here are fairly obvious if you look carefully enough at the diagrams.
b) (i) This one needs a bit of thought. There are certainly other good answers, and any answer that makes sense will be rewarded. For example, some very fine soil particles could escape the filter and affect the weight measurement.
c) The information tells us that the tray without grass (Tray A) loses more soil than the tray with grass (Tray B). So, the ***conclusion*** *is that the growing grass cuts down erosion.*

Making a prediction

To make predictions, you not only have to be able to read a graph or table accurately, you must also be able to predict what will happen under some other conditions. For example, these conditions could be increased temperature or allowing the experiment to run for longer.

Credit question 13

The graph shows the effect of temperature on the enzyme catalase.

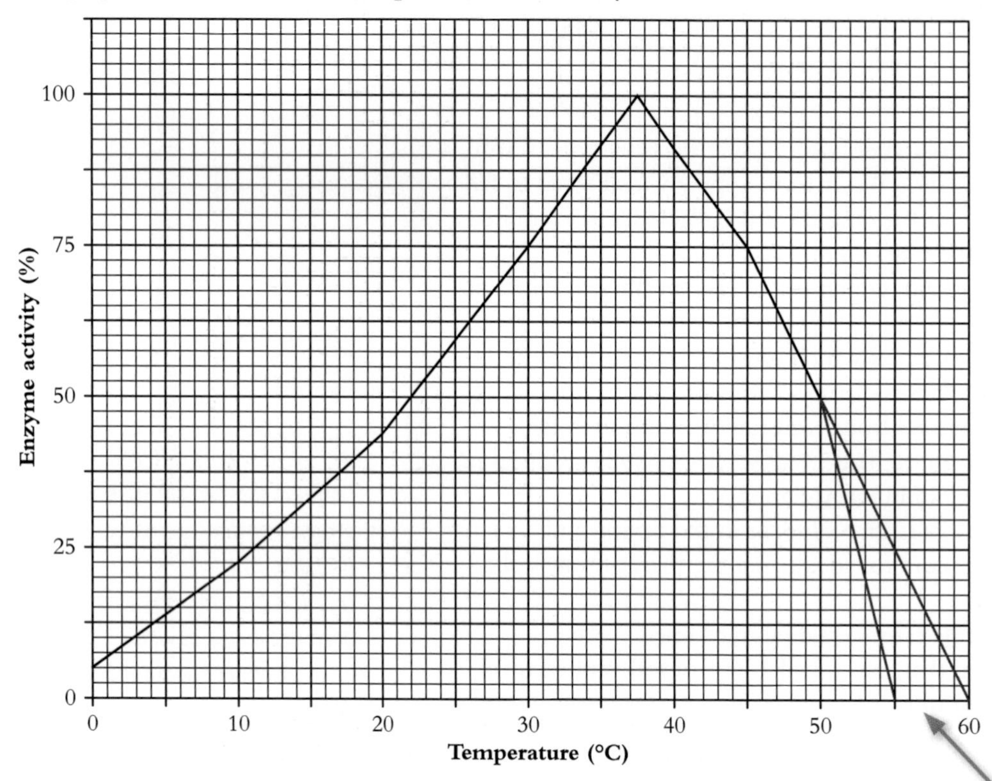

From the graph, predict the temperature at which the enzyme activity will reach zero. (1 mark)

Credit question 13 – Answer

Any temperature from 55°C to 60°C

A straight line continuing the slope of the graph reaches the x axis at 60°C, so that is one possible answer.

You could also calculate it as follows. In the 5°C between 45°C and 50°C, the rate of reaction drops by 25%. This means that it would take another 10°C rise to complete the final 50% drop in reaction to zero. 10°C above 50°C takes the temperature to 60°C.
It looks, however, as if the slope of the graph is getting steeper towards the end, so a value of between 55°C and 60°C would be acceptable.

Credit question 14

In a commercial process, a bacterial species is provided with glucose and produces a hormone. The bacteria release the hormone into the surrounding liquid. The graph shows the changes in the glucose concentration and the hormone concentration during a 60 hour period.

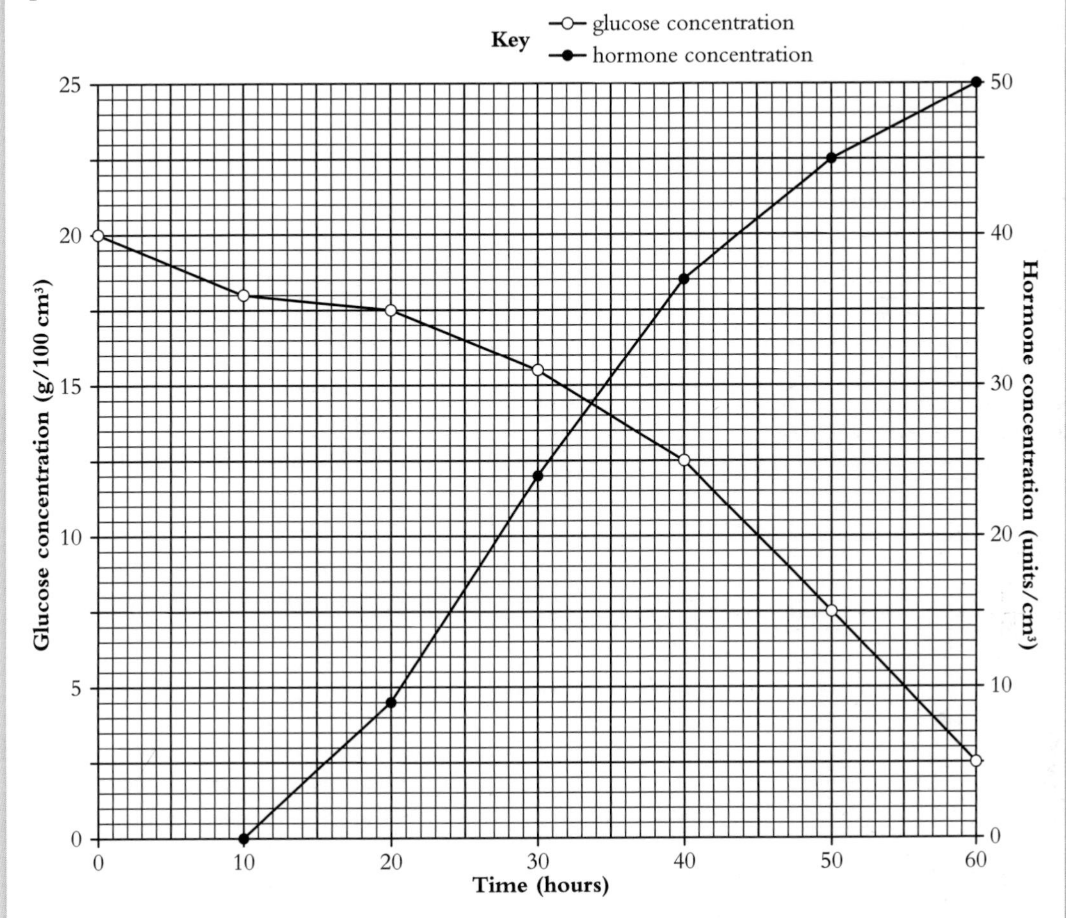

If the glucose concentration was used up at the same rate as between 50 and 60 hours, predict how many more hours it would be before all the glucose would be used up. (1)

Credit question 14 – Answer

5 hours

This answer needs to be calculated because we are at the end of the graph and cannot draw a continuing line. Be careful. On this graph there are two scales plotted on the vertical axes. Make sure you are reading the correct one. This question is asking about the glucose concentration so the appropriate scale is the one on the left.

As always, the starting point is what you are told in the question. Here you should be looking at the rate at which glucose is used up between 50 and 60 hours. In this time the concentration has gone down by 5g/100 cm³ to reach 2·5g/100 cm³.

If this drop of 5g/100 cm³ took 10 hours, then the remaining 2·5g/100 cm³ would take another five hours to reach zero.

Describing a relationship

A common type of problem solving question is to describe the relationship between two factors. Usually the information involved is in a graph, but it could relate to a table of results.

Credit question 15

The graph below shows the average length of young trout supplied with different concentrations of the element boron.

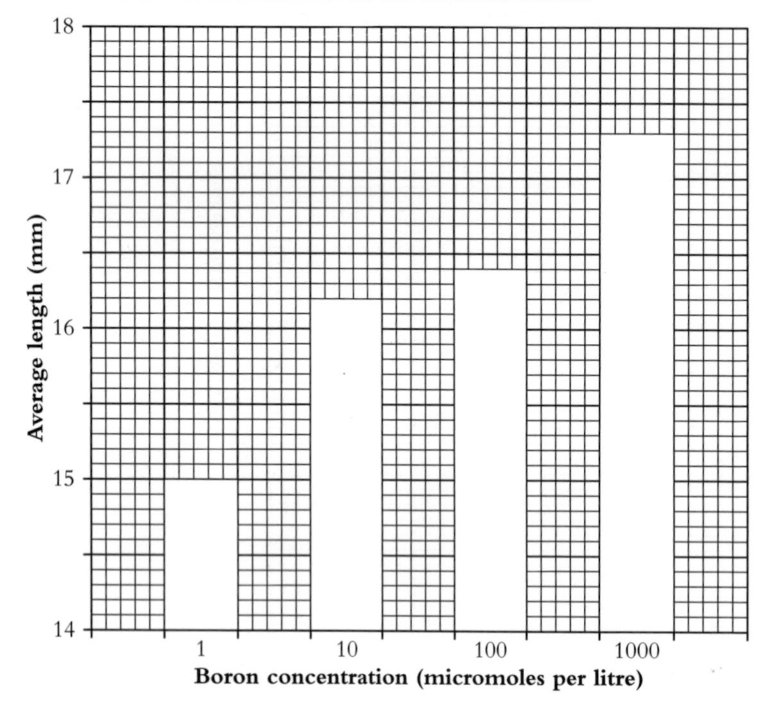

Describe the relationship between boron concentration and the length of the young trout. (1)

Credit question 15 – Answer

As the boron concentration increases, the average length of the young trout increases.

A common mistake is to get cause and effect mixed up when answering questions based on graphs or tables.
The answer must show that it is the change in boron which is causing the change in the trout. If you were to write, 'As the trout get bigger, the boron concentration increases', it would suggest that the trout were affecting the concentration of the boron, which is clearly wrong. You must put the factor which is causing the change first in your answer. If the question is based on a graph, you can be sure that the cause is the factor shown on the horizontal (x) axis. In this type of question, you must also make sure that you let the marker know exactly what you are talking about. There are two factors being considered here: the boron concentration and the length of the trout. If you start off your answer with, 'As it....', then it might not be clear which of the two is being described. Be safe and name the factors rather than using 'it'. Similarly, saying, 'As one increases the other increases', also tells us nothing about cause and effect.

Credit question 16

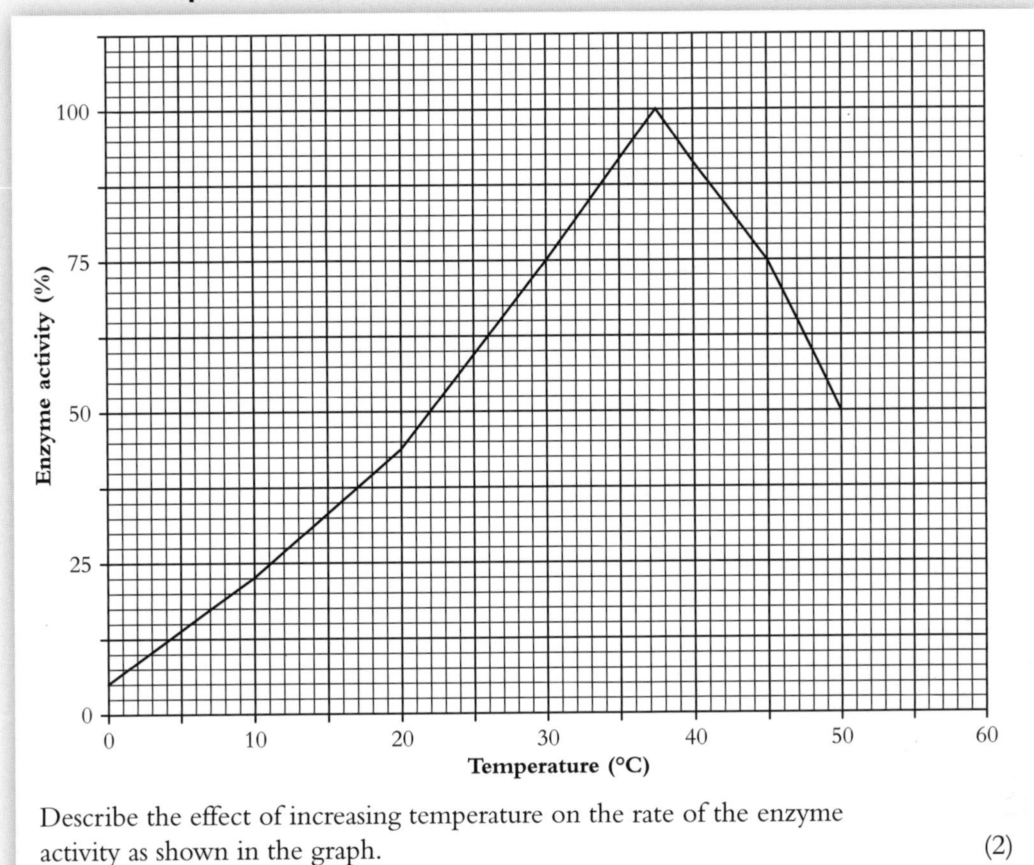

Describe the effect of increasing temperature on the rate of the enzyme activity as shown in the graph. (2)

Credit question 16 – Answer

As the temperature increases, the enzyme activity increases until it reaches a maximum at 37·5°C. Above this, the activity decreases as the temperature rises.

It is not enough to say, 'As the temperature rises, the enzyme activity increases.' This is only true until 37·5°C and so there would be no marks for this answer.
'The enzyme activity rises and then falls again', is a bit better, but you have not given any information about the temperature at which the change happens.
'It rises until 37·5°C' is still not enough, as you have not said what happens above 37·5°C. In both these cases you would lose marks.
Quoting 37·5°C shows that you have understood the scale and have read the graph correctly.
Adding the units °C in your answer makes sure of the mark.

Credit question 17

The table shows the volume of gas produced in one hour at different temperatures.

Temperature (°C)	10	20	30	40	50
Volume of gas produced in one hour (cm³)	9	18	36	48	5

Describe the relationship between the temperature and the volume of gas produced. (2)

Credit question 17 – Answer

As the temperature increases up to 40°C, the volume increases. Above 40°C, the volume decreases.

This looks simple – but it is all too easy to find yourself writing 'As the temperature rose to 40°C, the volume increased. At 50°C, the volume dropped.' Since there are no values between 40°C and 50°C, it is tempting to think that the volume did not drop until 50°C. If the results had been in a graph instead of a table, there would be no problem, as the line would clearly start dropping after 40°C.

So if you find that you cannot decide where the turning point is, try to think of the results as a graph. You will find that this takes away any uncertainty. In any graph, chart or table where you are asked to describe the relationship between factors, you must always quote the figure where the plot line changes direction.

Glossary

1 The Biosphere

abiotic factors non-living factors that can affect living organisms: for example, light, temperature, oxygen concentration
biomass the total mass of living organisms in a given area
community all the living organisms, both plants and animals, present in an ecosystem
competition the effect resulting from different organisms needing the same resource – one that is in limited supply
consumer an organism which gets its energy by feeding on other organisms or their wastes
denitrifying bacteria soil bacteria which break down nitrates in the soil and release nitrogen gas to the atmosphere
ecosystem all the members of a community, together with their habitats
food chain a series of feeding relationships between organisms where energy is passed on from one organism to the next; it must begin with a green plant
food web a complex pattern of connected food chains involving all the organisms in a community
fossil fuels fuels - including coal, oil and gas - which are formed from the remains of dead organisms; they are burned for heat and to generate electricity, and are a source of air pollution
habitat the area in which an organism lives; it provides particular conditions to which the organism is adapted
indicator species organisms whose presence - or absence - gives information about environmental conditions, such as levels of pollution
nitrifying bacteria soil bacteria which convert ammonia to nitrites, or nitrites to nitrates, in the soil
nitrogen fixing bacteria bacteria present in the soil or in the root nodules of some plants; they convert atmospheric nitrogen to nitrates
nuclear fuels radioactive materials which are used to generate electricity in nuclear power stations; the radioactive waste from this process must be stored securely for many generations
nutrient cycles processes which continually recycle elements such as carbon and nitrogen between living organisms and the environment
pitfall trap a method of collecting and sampling small invertebrates living on the soil surface
pollution the presence of harmful substances in the environment
population all the members of a single species in a given area
predator an animal which catches and kills another animal for food
prey an animal which is killed and eaten by another animal
producer green plants which make their food by photosynthesis; they are the first organisms of food chains
pyramid of biomass a diagram showing the total biomass present at each stage of a food chain
pyramid of numbers a diagram showing the total number of organisms present at each stage of a food chain
quadrat a method of sampling plants or stationary animals living in a given area
sewage organic waste from houses and other buildings that is washed into the sewers and treated at sewage treatment works

2 The World of Plants

animal external dispersal the spreading of seeds away from a parent plant on the outside of animals; the fruits have hooks which catch on the animals coats
animal internal dispersal the spreading of seeds away from a parent plant due to animals eating the fruits; the seeds pass from the animals in their faeces
anther the part of the stamen of a flower where pollen grains are produced and released
artificial propagation increasing the numbers of particular varieties of plants by any artificial method of asexual reproduction
asexual reproduction reproduction which does not involve fertilisation; only one parent is involved and there is no variation among the offspring
cellulose a carbohydrate made from glucose; it forms plant cell walls and so helps support plant structure
chlorophyll the green chemical which absorbs light energy needed for photosynthesis; it is present in cell structures called chloroplasts
chloroplasts structures present in some plant cells; they contain chlorophyll and are the sites of photosynthesis
clone a group of offspring which has been produced asexually; all the individuals in the group will be genetically identical to the parent
companion cell small cells found next to phloem vessels; they help control the activities of the phloem cells since these do not have nuclei
cuttings an artificial method of asexual reproduction in plants; small sections of stem are cut from a plant and encouraged to form roots and leaves
embryo the part of a seed which develops and grows into a new plant
epidermis the outer layer of cells of an organism; the lower epidermis of a plant leaf is where the stomata are found
fertilisation the stage of sexual reproduction during which the nucleus of a male gamete joins with the nucleus of a female gamete
flowering the stage of plant sexual reproduction during which the reproductive structures develop
food store material stored in the seed for use by the embryo during germination and early growth
fruit formation the stage of plant sexual reproduction during which the ovary containing developing seeds develops into a fruit
gamete a specialised cell used in sexual reproduction; the nucleus of a male gamete joins with the nucleus of a female gamete at fertilisation

germination the early stages of growth of the embryo in a seed; it includes the development of a root and a shoot
glucose the product of photosynthesis; it is used by the plant as food or to make structural materials, and can be stored as starch for later use
grafting an artificial method of asexual reproduction in plants: a cutting of one plant is joined to the stem of a stronger variety
guard cells two cells which surround a stoma and which control the opening and closing of it
insect pollination the transfer of pollen from anther to stigma by an insect
leaf veins transporting tissue in plant leaves; each vein contains both xylem and phloem vessels
lignin the chemical which strengthens the dead xylem vessels; the strong xylem vessels help support the plant
limiting factor a factor which controls how fast a reaction can take place or how large a population can grow, by not being present at a high enough level to allow any further increase
malting the controlled germination of barley grains so that stored starch is converted to maltose for fermentation
nectary small structures at the base of the petals of insect pollinated flowers; they produce nectar to help attract insects for pollination
ovary the female part of a flower in which the ovules are produced
ovule structures inside the ovary; an ovule contains the female gamete of the plant
palisade mesophyll the layer of cells in a leaf below the upper epidermis; these cells carry out most of the photosynthesis in the leaf
petal the coloured parts of a flower which help attract insects for pollination; they are not noticeable in wind pollinated flowers
phloem transport tissue in plants which carries sugars made in the leaves to storage or growing areas
photosynthesis the manufacture of glucose from carbon dioxide and water using absorbed light energy; oxygen is also produced
pollen grain small structures released from the anthers; a pollen grain contains the male gamete of the plant
pollination the stage of plant sexual reproduction during which pollen grains are transferred from an anther to a stigma
runners a natural method of asexual reproduction in plants, involving spreading horizontal shoots from which new plants develop
seed coat the outer layer of a seed which protects the internal embryo and food store
seed formation the stage of plant sexual reproduction during which the fertilised ovule develops into a seed within the ovary
sepal Small, leaf-like structures which protect a flower bud before it opens
sexual reproduction reproduction involving fertilisation (the joining of two specialised reproductive cells); it produces variation among the offspring
sieve plates perforated end cell walls which link phloem cells end to end; they help the movement of foods along the phloem vessels
spongy mesophyll the layer of cells below the palisade mesophyll; this layer has lots of air spaces between the cells and is where gas exchange happens
stamen the male part of a flower; the anthers where pollen grains are formed are found at the end of the stamens
starch a food substance made from glucose; it is used by plants as a food store
stigma the part of a flower which traps pollen grains during pollination
stoma /stomata (plural) the small openings or pores found on the bottom surface of leaves; carbon dioxide and oxygen can pass in and out through them, and water vapour is lost through them
tubers swollen underground stems which store food in some plant species; they allow asexual reproduction by sprouting to form new plants
wind dispersal the spreading of seeds away from a parent plant by the action of the wind; the fruits show adaptations to make this easier: for example, by developing wings
wind pollination the transfer of pollen from anther to stigma by the wind
xylem transport tissue in plants which carries water and minerals upwards from the roots; it consists of dead hollow tubes or vessels

3 Animal Survival

ADH the hormone which helps to control the water balance of the body by altering the volume of urine produced
amniotic sac a fluid-filled bag which protects the developing fetus from bumps
amylase an enzyme which digests starch into the sugar maltose; amylase is made in the salivary glands and the pancreas
anus the opening through which faeces is passed out of the body
bile a substance produced in the liver which helps fat digestion by breaking them down into tiny particles, with an increased surface area for enzyme action
bladder the organ which stores the urine from the kidneys until it is passed from the body
Bowman's capsule the part of a nephron which collects fluid filtered through the glomerulus
canines the teeth used by carnivores to kill, grip and tear food; they are gripping teeth in omnivores but may be absent in herbivores
carnivores animals that eat only animal materials; tigers are carnivores
collecting duct the structures which the kidney tubules lead to and which collect the urine
digestion the breakdown of large insoluble food molecules into small soluble products
eggs the female gametes of animals
embryo the early stage in the development of a young animal following fertilisation
external fertilisation when the sperm and egg meet outside the body of the female; this is needs to take place in water and is common in fish
fertilisation the joining of the nucleus of a male gamete with the nucleus of the female gamete
fetus the later stage in the development of a young animal; it is a growth stage and in humans it lasts from about eight weeks until birth at 40 weeks

filtration the first stage in the action of the kidneys; small molecules are filtered from the blood
gall bladder the gall bladder stores the bile from the liver and adds it to food in the small intestine
gametes specialised reproductive cells; one gamete from each parent join together at fertilisation
glomerulus a network a blood capillaries inside the Bowman's capsule which filters the blood
herbivores animals that eat only plant materials; sheep are herbivores
incisors the front teeth of mammals; they have a cutting action in herbivores and omnivores whilst in carnivores they grip and scrape food from bones
internal fertilisation the meeting of sperm and egg inside the body of the female; it happens with land-living animals such as mammals and birds
kidneys the organs used for water balance and for the removal of wastes such as urea from the blood
lacteal a small vessel in a villus; it absorbs the products of fat digestion
large intestine the part of the alimentary canal where water is absorbed from undigested food; the remaining material is faeces
lipase an enzyme which digests fats into fatty acids and glycerol; lipase is made in the pancreas
liver the organ which deals with the absorbed digestion products; it produces urea from excess protein and also produces bile
molars the back teeth are used to grind food in herbivores and to chew food in omnivores; in carnivores they are modified for slicing flesh
mouth the start of the alimentary canal where the teeth act on the food and saliva is added to it
nephrons the microscopic units in the kidneys where filtration and reabsorption take place
oesophagus the tube which connects the mouth to the stomach; food is pushed along it by peristalsis
omnivores animals that eat a mixture of plant and animal materials; humans are omnivores
ovaries the structures which produce eggs
oviduct a tube connecting an ovary with the uterus; it is where fertilisation takes place in mammals
pancreas an organ which makes digestive enzymes; these are added to food in the small intestine. The pancreas also produces the hormone insulin
peristalsis the muscular movements which push food along the alimentary canal; it consists of wave-like contractions of circular muscles
placenta a structure formed on the wall of the uterus where the blood systems of the fetus and the mother are close enough to allow exchange of materials between them
protease enzymes which digest protein: for example, pepsin in the stomach digests protein into peptides (short chains of amino acids)
reabsorption the second stage in the action of the kidneys; useful molecules are reabsorbed from the tubules back into the blood
rectum the part of the alimentary canal where faeces are stored before being passed from the body
renal artery the blood vessel which carries blood to the kidneys for filtration
renal vein the blood vessel which carries the filtered blood from the kidneys
response the reaction of an organism to a stimulus
rhythmical behaviour patterns of behaviour which happen repeatedly at regular intervals; they may be annual, daily or twice a day
salivary glands these glands in the mouth produce saliva which contains amylase enzyme
small intestine the part of the alimentary canal where digestion of food is completed and the absorption of digestion products into the blood takes place
sperm the male gametes of animals
stimulus any factor detected by an organism which produces a reaction
stomach a muscular part of the alimentary canal where food is held and mixed with acid and pepsin enzyme
testes the structures which produce sperm cells
trigger stimulus the stimulus which produces the start of rhythmical behaviour. It may be a change in daylength (annual), light / dark (daily) or tides (twice a day)
tubules the part of a nephron which the filtrate passes along and from which glucose and some water are reabsorbed
umbilical cord a tube containing blood vessels of the fetus and connecting the fetus with the placenta
urea a toxic waste material produced in the liver from the breakdown of excess amino acids
ureters tubes carrying the urine from the kidneys to the bladder
urine the liquid produced by the kidneys consisting of wastes, including urea, dissolved in water
uterus the part of the female reproductive system where a fertilised egg becomes attached and where the fetus develops
villi small projections on the inner lining of the small intestine; they create an increased surface area to speed up absorption
yolk a food store in an egg; in fish it is needed until the young fish can feed itself, whilst in birds it is needed until the young bird hatches

4 Investigating Cells

aerobic respiration the release of energy from foods such as glucose using oxygen; the products are carbon dioxide and water
catalase an enzyme which breaks down hydrogen peroxide into water and oxygen; it is present in all cells and has an optimum pH of 7
catalyst a substance which speeds up a chemical reaction but which is left unaltered after the reaction
cell the basic unit of living organisms; some organisms consist of just one cell and others consist of many millions of cells
cell membrane the structure which surrounds the cytoplasm in cells and controls the movement of materials into and out of the cell
cell wall the outermost layer of plant cells; it is made of cellulose fibres and provides support for the cells. It also prevents plants cells bursting if they are placed in pure water
chloroplast a structure which contains chlorophyll; chloroplasts are the sites of photosynthesis and are found in the cytoplasm of most leaf cells.

chromatid one of a pair of identical strands which together form a chromosome; they separate during mitosis and pass to different daughter cells
chromosomes structure in the cell nucleus; chromosomes contain the genetic information which controls cell activities and are visible during mitosis
chromosome complement the number and type of chromosomes present in a cell; mitosis keeps the chromosome complement the same from parent to daughter cells
concentration gradient the difference in concentration of a substance between two areas; the substance diffuses down the gradient from the high to the low concentration
cytoplasm jelly-like material in which the chemical reactions of the cell take place
denature permanent damage to an enzyme which prevents it from working; it is caused by high temperatures
diffusion the movement of a substance from an area of high concentration to an area of lower concentration
enzyme a biological catalyst which speeds up a cell reaction without being altered after the reaction
mitosis the division of a nucleus during cell division; the two daughter cells have the same chromosome complement as that of the parent cell
nucleus a structure in the cytoplasm of a cell which contains the genetic information controlling cell development and activities; it is surrounded by a membrane
optimum the condition in which an enzyme is most active; an enzyme will have an optimum temperature and an optimum pH for its activity
osmosis a type of diffusion in which water molecules move from a high water concentration to a low water concentration through a semi-permeable membrane
pepsin an enzyme which breaks down protein into peptides; it is produced by the stomach and works best in acidic conditions at pH 2·5
phosphorylase an enzyme which builds up starch from glucose-phosphate; it is found in some plant tissues
plasmolysed the condition of a plant cell which has lost water by osmosis; the vacuole shrinks pulling the cytoplasm and membrane away from the cell wall
product a substance which is formed during an enzyme controlled reaction: for example, the products of catalase on hydrogen peroxide are water and oxygen
selectively permeable the feature of cell membranes which allows small molecules to pass through but prevents the passage of larger molecules
specific the feature of an enzyme which means that it will only act on one particular substrate: for example, pepsin is specific for protein
spindle fibres a set of fibres to which the chromosomes become attached during mitosis
stain a chemical added to cell samples, to make them easier to see with a microscope
substrate a substance on which an enzyme acts: for example, the substrate for catalase is hydrogen peroxide
turgid the condition of a plant cell which contains as much water as it can: the pressure produced in the cell makes it firm
vacuole a fluid-filled sac found in plant cells; it pushes the cytoplasm and cell membrane against the cell wall, creating pressure which keeps the cell firm

5 The Body in Action

aerobic respiration the release of energy from foods such as glucose using oxygen, of which the products are carbon dioxide and water; this is the most efficient form of respiration
alveoli / air sacs microscopic structures at the ends of the bronchioles in the lungs; they are the sites of gas exchange between the air and the blood
anaerobic respiration the release of energy from food without the use of oxygen; it happens in muscle cells when there is an inadequate supply of oxygen
arteries blood vessels carrying blood away from the heart
atria / atrium (singular) the top two chambers of the heart which receive blood from the lungs (left atrium) and from the rest of the body (right atrium)
auditory nerve the nerve which carries signals from the ear to the brain
ball and socket joint a joint between bones which allows movement in several planes: for example, the hip and shoulder joints
binocular vision the use of two eyes to produce a three-dimensional image
bone structural tissue made of hard minerals and flexible fibres
brain the main part of the CNS (central nervous system) contained in the skull; it controls voluntary movements, thinking, memory and many other activities
breathing the action of taking air into the lungs (inhaling) and expelling air from the lungs (exhaling)
breathing rate the frequency of breathing measured as breaths taken per minute
bronchioles small branches from the bronchi which become microscopic air tubes ending at the alveoli
bronchus / bronchi (plural) the two branches of the trachea which lead into the lungs
cartilage smooth tissue which protects the ends of bones in a joint and which reduces friction; cartilage also strengthens the trachea and bronchi in the lungs
central nervous system (CNS) the part of the nervous system which receives sensory information and initiates appropriate responses. It consists of the brain and the spinal cord
cerebellum the part of the brain which co-ordinates body movements and balance
cerebrum the part of the brain which controls voluntary movements , thinking and memory
cilia hair-like structures which line the trachea and bronchi; they move with a wave-like action to sweep the mucus away from the lungs
cochlea the structure in the ear which converts sound vibrations into nerve signals
cornea the transparent front layer of the eye which allows light to enter
coronary arteries the arteries which carry blood into the muscle of the heart and so provide it with food and oxygen
diaphragm a muscular structure below the lungs which contracts and relaxes to change the volume of the chest during breathing
ears the organs of hearing which produce nerve signals from sounds; they also act as the organs of balance by producing nerve signals from movements of the head
ear drum the structure in the ear which vibrates as a result of sound waves reaching it

eyes the organs of sight which produce nerve impulses from images formed in them
fatigue the condition in which muscles are unable to work efficiently because of an accumulation of lactic acid: this results from anaerobic respiration
haemoglobin a red chemical present in the red blood cells; it combines with oxygen to form oxyhaemogloblin as the blood passes through the lungs, and is converted back to haemoglobin when oxygen is released at respiring tissues
heart the organ in the circulatory system which pumps the blood
hinge joint a joint between bones which allows movement in only one plane: for example, the knee and elbow joints
intercostal muscles the muscles between the ribs which produces the movement of the rib cage during breathing
iris the circular muscle in the eye which alters the size of the pupil to control the amount of light entering
lactic acid the product of anaerobic respiration which happens in muscles when the supply of oxygen is inadequate
lens the transparent structure which focuses the light into an image on the retina
ligament structures which hold the bones together in a joint
lungs the organs of gas exchange
medulla the part of the brain which connects with the spinal cord; it co-ordinates heart rate and breathing rate
middle ear bones small bones in the ear which transmit the vibrations of the ear drum to the cochlea
motor nerve cell a nerve cell which carries a signal from the CNS to a muscle or other effector
mucus a sticky substance produced in the trachea and bronchi. It traps bacteria and small particles of dirt, preventing them entering the lungs
muscle tissue which contracts to cause the movement of the bones in a joint
nerves structures which carry signals to the CNS from the sense organs and from the CNS to muscles and other effectors
nervous system the system which receives information from the senses and which co-ordinates the responses of the body; it consists of the CNS and the nerves which run to and from it
optic nerve the nerve which carries the signals from the eye to the brain
oxyhaemoglobin the chemical which transports the oxygen in the blood; it releases the oxygen as the blood passes through the body tissues
plasma the liquid part of the blood which transports dissolved substances such as glucose, urea, carbon dioxide and hormones; it carries the blood cells through the body
pulse a surge of blood through the arteries which can be felt where an artery is close to the skin surface; caused by the beating of the heart
pulse rate the frequency of pulse measured as beats per minute; this is equivalent to the heart rate
recovery time the time taken for the breathing rate and pulse rate to return to normal after exercise
red blood cells blood cells which transport oxygen from the lungs to the body cells and contain haemoglobin
reflex action a simple system in which a response to a stimulus is produced without the conscious involvement of the brain: for example, blinking
reflex arc a simple nervous pathway involving a sensory, a relay and a motor nerve cell
relay nerve cell a nerve cell within the CNS which provides links between other nerve cells
response an action taken as the result of specific information detected by the sense organs
retina the inner layer at the back of the eye where the image is formed and which produces the nerve signals from that image
semi-circular canals three structures in each ear which produce nerve signals from movements of the head
sensory nerve cell a nerve cell which carries a signal from a sense organ to the CNS
skeleton the system of jointed bones which supports the body, allows movement and protects some vital organs
spinal cord the part of the CNS contained in the spine; it co-ordinates some reflex actions and helps transmit signals between the brain and the rest of the body
stimulus information detected by the sense organs and which results in a particular response
synovial fluid liquid which lubricates a joint to reduce friction
synovial membrane a structure in a joint which produces synovial fluid
tendon fibrous tissue which connect muscle to bones
trachea the windpipe which leads from the throat to the bronchi
valves structures in the heart which prevent the backflow of blood; they are found between the atria and ventricles and also at the beginning of the arteries.
veins blood vessels carrying blood towards the heart
ventricles the bottom two chambers of the heart which pump blood to the lungs (right ventricle) and to the rest of the body (left ventricle)

6 Inheritance

alleles the different forms of a single gene: for example, the dominant and recessive form of a gene
amniocentesis the removal of a sample of amniotic fluid containing cells from a fetus and the examination of the cells for signs of chromosomal mutations
chromosomes structures in the nucleus of a cell which carry genetic information: body cells contain two complete sets of chromosomes but gametes contain only one set
continuous variation variation which shows a continuous range of possibilities between a minimum and a maximum value: for example, height, mass, handspan
discontinuous variation variation which shows a number of distinct groups rather than a range of possibilities: for example, blood group, eye colour, tongue-rolling ability
dominant the form of a gene which always shows its effect when present in an organism
F1 the symbol used for the first generation of offspring in a genetic cross
F2 the symbol used for the second generation of offspring in a genetic cross
gametes the sex cells of an organism; they contain one complete set of chromosomes rather than two sets, and so contain one gene for each characteristic rather than two

gene a unit of genetic information which is part of a chromosome: a gene controls one characteristic of an organism
genotype the genetic information an organism possesses for one characteristic
monohybrid cross a genetic cross in which the inheritance of only one characteristic is studied
mutagenic agent a factor which increases the chance of a mutation occurring: examples are X-rays, certain chemicals and heat shock
mutation a change to the number or structure of the chromosomes; such changes can be inherited and can affect the development of an individual
P the symbol used for the parental generation of a genetic cross
phenotype the appearance of a characteristic which results from the genotype possessed by an organism
recessive the form of a gene which is hidden when present with the dominant form in an organism
selective breeding the alteration of the characteristics of a group of organisms over many generations, by choosing which individuals are allowed to breed because they possess desirable characteristics
species a group of interbreeding organisms whose offspring are fertile
true-breeding individuals which possess two identical alleles for a characteristic: either two dominant or two recessive alleles
variation differences which can be recognised between individuals of the same species
X and Y chromosomes the chromosomes which determine the sex of an individual

7 Biotechnology

alcohol a product of anaerobic respiration by yeast
anaerobic respiration the release of energy from food without using oxygen; less efficient than aerobic respiration
antibiotics chemicals, produced by certain micro-organisms, which kill or prevent the growth of other micro-organisms: they are important as medicines
bacteria a group of micro-organisms, some of which are beneficial and some of which are harmful
batch processing an industrial process in which the starting materials are mixed and the process allowed to continue to completion. the product must then be separated from the reaction mixture
biological detergents detergents, such as washing powders, which contain digestive enzymes to help remove stains at relatively low temperatures
biotechnology the use of micro-organisms to manufacture products useful to people
circular chromosome a type of chromosome found in many types of bacteria; there is one such chromosome in a bacterial cell
continuous flow process an industrial process in which starting materials are continually added and the product is continually removed without stopping the process
diabetes a condition in which the body is unable to control the blood sugar levels; it may be caused by lack of insulin and treated with injections of insulin
fermentation anaerobic respiration carried out by micro-organisms
fossil fuels fuels, including coal, oil and gas which are formed from the remains of dead organisms; they are burnt for heat and to generate electricity
genetic engineering the altering of the genetic material of micro-organisms to give them new characteristics such as the ability to make useful chemicals
immobilisation a technique in which enzyme molecules or microbial cells are fixed in place, for example to jelly beads, for use in continuous flow processes.
insulin the hormone needed for control of blood sugar levels; human insulin can be made by genetically modified bacteria and used to treat diabetes
lactose the sugar present in milk
malting the controlled germination of barley grains so that stored starch is converted to maltose for fermentation
maltose the sugar produced from the digestion of starch by amylase
methane the gas produced during the decomposition of organic material such as sewage in anaerobic conditions; it is can be used as a fuel
mycoprotein protein, produced by some fungi, which can be used a food source for people and animals
plasmids small circular chromosomes found in bacterial cells; they can be manipulated and used to introduce new genes into bacteria during genetic engineering
sewage organic waste from houses and other buildings that is washed into the sewers and treated at sewage treatment works
spores small reproductive structures formed by bacteria and fungi which can survive adverse conditions
upgraded waste material with an increased energy or protein content that has been produced by micro-organisms from organic waste
yeast a single-celled fungus which produces alcohol from sugar during anaerobic respiration: it is important in the manufacture of bread and alcoholic drinks

Problem Solving

control part of an experiment which proves that the results obtained are due to one particular factor
hypothesis a suggested explanation of why something happens; the hypothesis still must be proven by a suitable experiment
reliability a measure of the confidence you can have in a set of results; results can never be 100% reliable, but they can be made more reliable by repeating the experiment
validity a measure of how suitable an experiment is for its intended purpose; a valid experiment has a control in which only one factor changes